Compelled Compassion

Contemporary Issues
in Biomedicine, Ethics, and Society

Compelled Compassion: *Government Intervention in the Treatment of Critically Ill Newborns,* edited by *Arthur L. Caplan, Robert H. Blank, and Janna C. Merrick,* 1992

New Harvest: *Transplanting Body Parts and Reaping the Benefits,* edited by *C. Don Keyes,* 1991

Ethics and Aging, edited by *Nancy S. Jecker,* 1991

Beyond Baby M: *Ethical Issues in New Reproductive Techniques,* edited by *Dianne M. Bartels, Reinhard Priester, Dorothy E. Vawter, and Arthur L. Caplan,* 1989

Reproductive Laws for the 1990s, edited by *Sherrill Cohen and Nadine Taub,* 1989

The Nature of Clinical Ethics, edited by*Barry Hoffmaster, Benjamin Freedman, and Gwen Fraser,* 1988

What Is a Person?, edited by *Michael F. Goodman,* 1988

Advocacy in Health Care, edited by *Joan H. Marks,* 1986

Which Babies Shall Live?, edited by *Thomas H. Murray and Arthur L. Caplan,* 1985

Feeling Good and Doing Better, edited by *Thomas H. Murray, Willard Gaylin, and Ruth Macklin,* 1984

Ethics and Animals, edited by *Harlan B. Miller and William H. Williams,* 1983

Profits and Professions, edited by *Wade L. Robison, Michael S. Pritchard, and Joseph Ellin,* 1983

Visions of Women, edited by *Linda A. Bell,* 1983

Medical Genetics Casebook, by *Colleen Clements,* 1982

Who Decides?, edited by *Nora K. Bell,* 1982

The Custom-Made Child?, edited by *Helen B. Holmes, Betty B. Hoskins, and Michael Gross,* 1980

Medical Responsibility, edited by *Wade L. Robison and Michael S. Pritchard,* 1979

Contemporary Issues in Biomedical Ethics, edited by *John W. Davis, Barry Hoffmaster, and Sarah Shorten,* 1979

Compelled Compassion

*Government Intervention
in the Treatment of Critically Ill Newborns*

Edited by

Arthur L. Caplan,

*Center for Biomedical Ethics, University of Minnesota,
Minneapolis, Minnesota*

Robert H. Blank,

*University of Canterbury,
Christchurch, New Zealand*

and

Janna C. Merrick

*St. Cloud State University,
St. Cloud, Minnesota*

 Humana Press • Totowa, New Jersey

© 1992 The Humana Press Inc.
999 Riverview Dr., Suite 208
Totowa, NJ 07512

Printed in the United States of America

Library of Congress Cataloging in Publication Data

Main entry under title:

Compelled compassion ; government intervention in the treatment of
  critically ill newborns / edited by Arthur L. Caplan, Robert H.
Blank, and Janna C. Merrick.
    p. cm. — (Contemporary issues in biomedicine, ethics, and
society)
  Includes index.
  ISBN 0-89603-224-8
  1. Neonatal intensive care—Law and legislation—United States.
2. Neonatal intensive care—United States—Moral and ethical
aspects. I. Caplan, Arthur L. II. Blank, Robert H. III. Merrick,
Janna C. IV. Series.
RJ253.5.C65 1992
618.92`01—dc20   91-44190
                                                          CIP

Dedication

For Janna's parents, Les and Sadie Lowrey

Acknowledgments

The editors are indebted to a number of people. Carolyn Braun critiqued each chapter and was foolish enough to ask for more work. So we obliged and gave her more. Mary Howell assisted with a variety of administrative tasks and generally kept the operation running smoothly. Toni Knezevich spent endless hours, always with a smile on her face, editing manuscripts and preparing them for the publisher. Parks Walker lent both editorial and moral support. Tom Lanigan, Lucia Read, and Fran Lipton at Humana did a marvelous job of shepherding the book through production. We all owe an intellectual debt to the late Nancy Rhoden who contributed so much scholarship to the Baby Doe debate. We had originally invited her to contribute a chapter and were greatly saddened by her untimely death.

Preface

In April 1982, an infant boy was born in Bloomington, Indiana, with Down syndrome and a defective, but surgically correctable, esophagus. His parents refused to consent to surgery or intravenous feeding. The hospital unsuccessfully sought a court order to force treatment, and appeals to higher courts also failed. The child, identified as Baby Doe by the news media, subsequently died. The events in Bloomington became the catalyst for action by the Reagan administration, the courts, and Congress that culminated in a federal policy that makes failure to treat newborns with disabilities a form of child neglect.

This book centers on the public policy aspects of withholding treatment from critically ill newborns who are disabled. Specifically, it deals with why the policy was enacted and what impact it has had on health care workers, families, and infants. Some of the contributors to this book spearheaded the early debate on withholding treatment. Anthony Shaw's *New York Times Magazine* article in 1972 was the first to address these issues in the popular press. The following year, he published a related article in the *New England Journal of Medicine*. Also appearing in this same issue of *NEJM*, was the pathbreaking study, coauthored by A. G. M. Campbell, on withholding treatment in the special care nursery at Yale-New Haven Hospital. Each of these articles promoted much public and professional discussion.

Other contributors to this volume were active as participants in the policy making process itself. Norman Fost served on the American Academy of Pediatrics Committee on Bioethics, which played a critical role in developing federal policy. James Bopp, President of the National Legal Center for the Medically Dependent and Disabled and General Counsel for the National Right to Life Committee, played a role in negotiating the compromises that led to federal legislation. Rutherford Turnbull participated in drafting the "Principles of Treatment of Disabled Infants," a joint policy statement reflecting the views of nine medical and disability rights organizations that became the basis for the early regulations. All three have written numerous articles on Baby Doe issues in professional journals.

Treatment decisions for critically ill newborns are often made and implemented in an atmosphere of crisis. Clinicians may disagree about the diagnosis and prognosis for a given child. Parents may be in a state of shock and grief, complicated by a lack of understanding about their baby's condition. It is in this highly charged setting that federal legislation seeks to "compel compassion" by making failure to provide treatment in certain cases a form of child neglect. There are many who believe that medical staff and parents are filled with compassion and seek what is in the best interests of critically ill newborns. Others feel that such compassion is absent, and therefore must be compelled by law.

It has been a decade since the death of Baby Doe, a decade in which medical science has made tremendous advances in clinical technology. It is our hope that this book will contribute to current public and professional awareness of the complexity of treatment decisionmaking for newborns who are disabled and who are also critically ill.

Janna Merrick

Contents

vii Preface

xi Contributors

1 Life-and-Death Decisions in the Midst
   of Uncertainty
      Robert F. Weir

35 Conflict, Compromise, and Symbolism:
    The Politics of the Baby Doe Debate
       Janna C. Merrick

73 A Legal Analysis of the Child Abuse
    Amendments of 1984
       James Bopp, Jr. and Mary Nimz

105 Hard Cases Make Bad Law:
    The Legacy of the Baby Doe Controversy
       Arthur L. Caplan

123 Parental Perspectives on Treatment–Nontreatment
    Decisions Involving Newborns with Spina Bifida
       *Patricia A. Barber, Janet G. Marquis,
       and H. Rutherford Turnbull III*

155 Rationing Medicine
    in the Neonatal Intensive Care Unit (NICU)
       Robert H. Blank

185 Baby Doe and Me: *A Personal Journey*
       Anthony Shaw

207 Baby Doe and Forgoing Life-Sustaining Treatment:
   Compassion, Discrimination, or Medical Neglect?
   A. G. M. Campbell

237 Neonatologists, Pediatricians, and the Supreme Court
   Criticize the "Baby Doe" Regulations
   Loretta M. Kopelman, Arthur E. Kopelman,
   and Thomas G. Irons

267 The Impact of the Child Abuse Amendments
   on Nursing Staff and Their Care of Handicapped
   Newborns
   Joy Hinson Penticuff

285 Infant Care Review Committees in the Aftermath
   of Baby Doe
   Norman Fost

299 Decision Making in the Neonatal Intensive Care Unit:
   The Impact of the 1984 Child Abuse Amendments
   Terry Walman

317 Appendix:
   Chronology of Events Related to Passage
   of the 1984 Child Abuse Amendments

321 Biographies

329 Index

Contributors

PATRICIA BARBER • *University of Kansas, Lawrence, Kansas*

ROBERT H. BLANK • *University of Canterbury, Christchurch, New Zealand*

JAMES BOPP • *Brames, McCormick, Bopp, and Abel, Terre Haute, Indiana*

A. G. M. CAMPBELL • *University of Aberdeen, Aberdeen, Scotland*

ARTHUR L. CAPLAN • *University of Minnesota, Minneapolis, Minnesota*

NORMAN FOST • *University of Wisconsin Hospital and Clinics, Madison, Wisconsin*

THOMAS G. IRONS • *East Carolina University, School of Medicine, Greenville, North Carolina*

ARTHUR KOPELMAN • *East Carolina University, School of Medicine, Greenville, North Carolina*

LORETTA KOPELMAN • *East Carolina University, School of Medicine, Greenville, North Carolina*

JANET G. MARQUIS • *University of Kansas, Lawrence, Kansas*

JANNA C. MERRICK • *St. Cloud State University, St. Cloud, Minnesota*

MARY NIMZ • *National Legal Center for the Medically Dependent and Disabled, Inc., Indianapolis, Indiana*

JOY HINSON PENTICUFF • *University of Texas, Austin, Texas*

ANTHONY SHAW • *University of California at Los Angeles Medical Center, Pasadena, California*

H. RUTHERFORD TURNBULL • *University of Kansas, Lawrence, Kansas*

TERRY WALMAN • *Johns Hopkins University School of Medicine, Baltimore, Maryland*

ROBERT WEIR • *University of Iowa, College of Medicine, Iowa City, Iowa*

Life-and-Death Decisions in the Midst of Uncertainty

Robert F. Weir

A Neonatal Intensive Care Unit (NICU) is characterized by premature and disabled patients with life-threatening conditions, highly trained medical and nursing specialists, state-of-the-art medical technology, an endless stream of medical consultants, parents grappling with frightening possibilities, and numerous decisions that have to be made in the midst of impenetrable uncertainty. Whether made while looking down at an imperiled baby, in consultation with the baby's parents, or in a conference room near the NICU, many of these decisions are crucial because a baby will continue to live or will die as a consequence of the decisions.

The context of uncertainty has several parts. The primary uncertainty in many NICU cases is medical in nature, with the greatest medical uncertainty usually having to do with prognosis rather than diagnosis. Even with the availability of increasingly sophisticated diagnostic tools, neonatologists and other pediatric specialists sometimes find it impossible, literally impossible, to predict with any measure of confidence the long-term prospects of a child with severe neurologic, cardiac, genetic, or pulmonary disabilities. Such

From: *Compelled Compassion* Eds.: Caplan, Blank, and Merrick
©1992 The Humana Press Inc.

medical uncertainty most commonly arises with babies having known neurologic impairments, but with pediatric neurologists and developmental specialists having no way of
accurately predicting the extent or severity of the developmental disability that the child will actually have if he or
she lives.

The other parts of the context of uncertainty in the truly
difficult NICU cases have to do with uncertainties pertain-
ing to ethics, the law, and the parents of the baby. Moral
uncertainty arises for either of two reasons: The neonatologists, parents, and other participants in a case are not adequately informed about the work that has been done in
neonatal ethics in recent years, or the uncertainty about the
medical prognosis in a given case necessarily causes the ethical analysis of the case to be less decisive than it would
normally be.

Uncertainty about the law takes many forms. Sometimes
the decision makers in a case are insufficiently informed
about how federal or state statutes apply to the case under
consideration. Sometimes they are unsure regarding the
meaning and applicability of the federal child abuse regulations and about the role that can or should be played by state
child protective agencies in cases involving decisions by parents not to sustain the lives of severely disabled newborns.
In some cases, the legal uncertainty is focused on a particular question, such as the legal liability of physicians who
refuse to abide by parents requests regarding the administration or abatement of treatment, the legality of transplanting organs from an anencephalic baby, or the legality of
abating life-sustaining nutrition and hydration.

Uncertainty about the parents of an imperiled baby adds
another level of complexity to some cases. At times neonatologists, NICU nurses, or consultants on particular cases
raise questions about the parents' abilities to grasp the relevant medical data, their psychological conditions, their
desires related to the baby's life or death, their abilities to
cope with the ongoing management problems presented by

their disabled child, and their abilities to pay for the hospital and medical expenses involved in the child's care. At other times, much more specific case-centered questions are raised, such as questions about the mother's use of alcohol, cocaine, and other drugs during pregnancy, the absence of an identifiable father, the mother's earlier refusal of prenatal diagnosis, parental decisions made contrary to the advice of genetic counselors, problematic interactions between parents, or the likelihood that the parents may initiate legal action if their requests regarding treatment or the abatement of treatment are not carried out.

The commonality of uncertainty in its several forms will be illustrated in two ways in this chapter. First, five clinical cases will be briefly presented. These cases have occurred in hospitals in four states. All of the cases have taken place since the "Baby Doe" regulations were first published in March 1983 and the subsequent Child Abuse regulations were published in April 1985. Second, five questions will be addressed. These questions focus on substantive aspects of decision making about neonates and other young infants that remain problematic for persons making life-and-death decisions at either the clinical or public-policy level.

Five Cases

Case 1

Baby girl A was born with myelomeningocele and hydrocephalus in a denominational hospital. She was quickly transferred to a level III NICU in a university teaching hospital in the same city. After doing a variety of diagnostic studies, the medical staff was concerned about the size of the myelomeningocele, the severity of the hydrocephalus, and the thinness of the neocortical mantle. Nevertheless, the physicians recommended surgery, at least partially because they believed that the recently published Child Abuse regulations left them no other legal alternative.

The baby's parents refused to consent to surgery. After extensive conversations with the physicians in the case, the parents believed that the loving and humane thing to do was to let their daughter die, rather than to subject her to a life with the kinds of severe neurological and physical disabilities the physicians said lay ahead of her even with multiple surgeries.

The case was referred to the hospital's Infant Care Review Committee (ICRC). The committee discussed the case with the attending neonatologist, as well as with a developmental disabilities specialist and the surgeon scheduled to do the operation. The committee also discussed the case with the parents, who brought the baby's grandmother and their minister to the meeting. After three hours of deliberation, the committee reluctantly decided to recommend the surgery—even though they knew the parents would continue to refuse consent, the hospital would be obligated to inform the state Department of Human Services, the DHS would likely initiate court action against the parents for child abuse, the baby's father could lose his job as a police officer in the city, and the baby would grow up severely retarded and disabled. The committee members were significantly influenced by the hospital's attorney, who argued that institutional adherence to the Child Abuse regulations was more important than ethical concerns based on the infant's best interests.

Case 2

Six-month-old Sammy Linares, having aspirated a deflated balloon and had a cardiac arrest, was rushed to the emergency room at MacNeal Hospital in Chicago following the removal of the balloon by paramedics. He was resuscitated and placed on an infant ventilator, even though he had been without spontaneous cardiac function for approximately 20 minutes. He was then transferred to the Pediatric Intensive Care Unit (PICU) at Rush-Presbyterian-St. Luke's Medical Center, where he again was placed on a ventilator.

The young boy was not expected to survive 24 hours. However, his cardiorespiratory status stabilized, and the prognosis for the comatose young patient changed to the expectation that he would survive—but with profound and permanent brain damage. After several months, he met the diagnostic criteria for persistent vegetative state.

Soon after Sammy's admission to Rush, his father requested that the ventilator sustaining his son's life be disconnected. The physicians in the case, following the advice of the hospital attorney, refused to act on that request. Four months later, Sammy's father took his son off the ventilator himself, but was restrained by security guards and the ventilator was reconnected. Eight months later, after continuously being told that disconnecting the ventilator would be contrary to Illinois law, he entered the PICU with a handgun, held the hospital staff at gunpoint, disconnected the ventilator, and let Sammy die in his arms (Gostin, 1989; Goldman et al., 1989).

Case 3

Baby girl L was born following a pregnancy complicated by fetal hydronephrosis and oligohydramnios in the last trimester. The baby was resuscitated after delivery, presented Apgar scores of 1, 4, and 5, and was placed on a ventilator. She subsequently stabilized and was weaned from the ventilator, but remained severely impaired neurologically. Over the next 23 months, she had a gastrostomy, a tracheostomy, numerous episodes of aspiration and uncontrolled seizures, pneumonia, septic shock, and four cardiopulmonary arrests.

When she was again hospitalized, the primary care physicians, chiefs of service, nurses, hospital counsel, and chairpersons of the institutional ethics committee agreed unanimously that reinstituting mechanical ventilation and cardiovascular support was contrary to the patient's best interests. However, the child's mother, having continuously demanded that everything possible be done to ensure her daughter's survival, rejected the recommendation to abate further life-sustaining interventions.

An unusual probate court hearing then took place. The mother's attorney argued for resuming mechanical ventilation for the child. The physicians and hospital attorney resisted that course of action, arguing that further medical intervention was not in the patient's best interests. The child was subsequently transferred to the care of a physician in another institution and, two years later, remains blind, deaf, quadriplegic, profoundly retarded, and dependent on intensive home nursing care (Paris et al., 1990).

Case 4

Baby boy G was transferred to the NICU of a teaching hospital with the diagnosis of severe hyperammonemia. This metabolic disorder, once universally fatal, is now sometimes treated in an experimental manner with a combination of dietary protein reduction, antibiotics, experimental medications to try to correct the urea cycle, peritoneal dialysis, and exchange transfusions. In this case, the parents initially consented to a combination of dialysis and experimental medications, even though they were informed of the severity of the biochemical problem and the likelihood that their son would be neurologically impaired if he survived.

The parents' first child had died from the same condition, and genetic counselors had advised them of the risks inherent in a second pregnancy. Nevertheless, they had initiated a second pregnancy, and then refused prenatal diagnosis and delivery at the teaching hospital. Two days after giving consent for treatment, they told the physicians they wanted the treatment stopped—even though the baby's medical condition was improving with the experimental treatment protocol.

The medical staff was surprised at the baby's improving condition and the parents' request to abate treatment. During a multidisciplinary case conference, the consensus view was that an ironic situation had developed: The parents had consented to treatment when there was great uncertainty about prognosis, but then refused to continue the treatment when medical intervention seemed more clearly in the baby's best interests. With great uncertainty and reluctance, the physicians acquiesced to the parents' decision, and the baby died a day later.

Case 5

Baby girl T was born at 32 weeks gestation, was intubated at birth, and presented numerous congenital anomalies. She was diagnosed at a university hospital as being microcephalic and having the VATER syndrome, which includes esophageal atresia, tracheoesophageal fistula, cardiovascular anomalies, and vertebral defects. During her time in the NICU, she had an esophagostomy, numerous cardiac catheterizations and operations, a gastrostomy, a tracheostomy, a grade II intraventricular hemorrhage, multiple infections, bronchopulmonary dysplasia, pulmonary hypertension, and multiple seizures.

Nevertheless, she survived and was subsequently transferred to a pediatric chronic care unit in the hospital. Her months there were characterized by ongoing severe cardiac, neurologic, and pulmonary problems. Pediatric cardiologists occasionally disagreed about the severity of her cardiac conditions. Pediatric pulmonologists continuously adapted their management of her brochopulmonary dysplasia as her medical condition changed because of infections and surgeries. Nurses suctioned and bagged her frequently when her tracheostomy became blocked. Medical specialists rotating through the unit disagreed about whether she should be placed on do-not-resuscitate status.

Because of her specialized medical and nursing care, the young girl survived—but did not leave the chronic care unit. When she reached 30 months of age, she had been in the hospital her entire life. Her mother, who lived in another city, visited the unit only a few times a year, and her father never came. Her mother could not take her home, and the social workers in the unit had been unable to arrange any other placement for the severely disabled young girl. The family's financial resources were quite limited, and the cost of the girl's medical care had by that time exceeded $2 million.

Five Questions

Do Neonates Count as Persons?

The first question of uncertainty has two dimensions. One dimension is theoretical, and has to do with the ontological status of neonates and young children, whether normal or neurologically impaired. Simply put, this aspect of the question asks, "What kind of entity is it whose life or death is at stake in the decisions made by clinicians and/or parents?" The other dimension is practical, and pertains to the realities of decision making in NICUs, PICUs, and specialized chronic care units for young children. This second aspect of the more general question asks, "What difference does an abstract question about personhood make in the real world of tough clinical decisions?"

The uncertainty surrounding the theoretical question of personhood could hardly be greater, especially among persons lacking training in philosophy. Although the debate about the ontological status of human beings in certain situations (both before and after birth) has been going on for centuries among philosophers and theologians, a question about the ontological status of neonates nevertheless elicits puzzled responses when raised in a modern medical center, university, or other social setting. The question often seems strange to health-care professionals and others outside philosophy, partially because of its abstract nature and partially because many people simply assume that "person" and "human" are synonyms.

Sometimes the abstractness of the question about personhood can be alleviated by suggesting to individuals or groups that they try to articulate what they understand the term "person" to mean, especially in terms of the properties that persons possess. Then, one can raise questions about the possession of those properties by patients in a persistent vegetative state, by severely demented individuals, by dolphins, by some apes, by extraterrestrial beings, or by human babies. The central point in such a discussion is clear, namely whether any of the human or nonhuman candidates for personhood are correctly to be regarded as persons in the same ways that you and I are persons.

On other occasions, the abstractness of the question about the personhood of neonates can be alleviated by pointing out that the language of personhood regularly surfaces in prolife arguments about the morality and legality of abortion. For example, one of the many proposed amendments to the US Constitution following the 1973 *Roe* v *Wade* decision would have granted fetuses the legal status of persons:

> With respect to the right to life, the word "person" as used in this article and in the fifth and fourteenth articles of amendment to the Constitution of the United States, applies to all human beings, including their unborn offspring at every stage of their development, irrespective of age, health, function, or condition of dependency (Perkins, 1974).

Thus, the question of personhood is with us, along with its abstractness, and its legitimate application covers the spectrum of human biological existence from fetuses to permanently unconscious adult patients to severely demented geriatric patients. The question of the personhood of neonates, at least among philosophers, has generally been framed so that it focuses on whether neonates possess the properties or intrinsic qualities generally ascribed to persons, such as you and I (Tooley, 1972,1979,1983; Warren, 1973; Singer, 1979). A few philosophers have suggested that the personhood of neonates, if they are to be counted as persons, can also include their social roles or extrinsic qualities (Engelhardt, 1986; Jecker, 1990).

To the extent that there is consensus among philosophers on the concept of personhood, that consensus focuses on the intrinsic rather than the extrinsic qualities of persons. Most philosophers agree on at least the core properties or traits of personhood, if not on all of their applications. Joel Feinberg (1986), in his discussion of "commonsense personhood," puts forth the consensus view of personhood as being the possession of three necessary and jointly sufficient properties:

consciousness, self-awareness, and at least minimum ratio-
nality. Such properties, for him and many others, represent
"person-making characteristics."

The possession of personhood, therefore, has to do with
neurological development and, at least among human beings,
the absence of profound neurological dysfunction or impairment. The answer to the question of whether neonates are to
be counted as persons depends on three interrelated factors:

1. How much neurological development is required for
   personhood;
2. How much neurological impairment is necessary to rule
   out personhood; and
3. Whether any significance is to be placed on the principle
   of potentiality as it applies to personhood.

In my judgment, there are three basic positions regarding the personhood of neonates (and other human beings
whose personhood may be questioned), and the positions are
distinguishable largely because of their handling of the factors of neurological development, neurological impairment,
and potential personhood (Weir, 1985, 1986). The first position holds that *all neonates*, whether normal or neurologically impaired, *are nonpersons*. As initially described by
Thomas Murray (1985), the philosophers who hold this view
make four related claims:

1. Personhood is a moral category attaching to beings (of any
   species) with certain characteristics, principally cognitive
   capacities;
2. Neonates lack the cognitive qualities that make a human
   into a person;
3. Being a *potential* person does not count; and
4. Only *actual* persons are entitled to the moral benefits of
   personhood, including the right not to be killed.

The second position stands at the other end of a philosophical and political spectrum, and represents a very common view of neonates held by many physicians, nurses, and

other people as well. This position, which holds that *all neo-nates are actual persons*, can also be presented in terms of four related claims:

1. Personhood is a moral and legal category attaching to human beings at a chronological point determined by social consensus, with conception, viability, and birth as three alternatives for such a point;
2. Neonates count as *actual* persons, as do all other human beings who are past the chosen chronological point;
3. Being a *potential* person does not signify very much; and
4. Neonates and all other *actual* persons are entitled to the moral and legal benefits of personhood, including the right not to be killed.

The third position stands between the other positions, differing from the first position's insufficient claims and the second position's excessive claims regarding the personhood of human newborns. This position, which holds that *most neonates are potential persons*, can be compared with the alternative views on the basis of its four claims:

1. Personhood is a moral category attaching to beings (of any species) with certain characteristics, principally cognitive capacities;
2. Neonates lack the intrinsic qualities that make a human into a person, as do fetuses;
3. Having the potential to become a person through the normal course of development does count, and neonates without severe neurological impairment (and fetuses having exhibited brain activity) have this potential; and
4. All *potential* persons have a *prima facie* claim to the moral benefits of personhood, including the right not to be killed, because they will subsequently acquire an *actual* person's moral and legal right to life.

The last of these positions, in my view, is the correct way of describing the ontological status of neonates. This position is preferable to the neonates-are-not-persons view of some philosophers, because it grants more than a species value to human newborns—and avoids the major weakness of having to allow, in principle, for the indiscriminate termination of an indeterminate number of neonatal lives, whether these lives are cognitively impaired, physically disabled, or normal. The third position is also preferable to the neonates-are-actual-persons view (especially as put forth by several prolife groups), because it takes the philosophical and psychological concept of personhood seriously—and avoids the major weakness of having to say, in principle, that a baby has no more claim to the moral benefits of personhood than an early human embryo does.

Now we turn to the second aspect of the question about personhood: What difference does greater clarity about the ontological status of neonates make in the real world? Given the abstract, nonempirical nature of the concept of potential personhood, how can such a concept—even if philosophically sound—be used to make the tough clinical decisions about initiating, continuing, or abating life-sustaining treatment in an NICU?

The honest answer is, of course, that the philosophical view developed above cannot currently play a major role in clinical decision making for a number of reasons. It is too abstract and subjective. It is not based on, nor is it likely to be based on, empirical studies. It is the sort of view about which reasonable people disagree. It opens up the possibility of using quality-of-life criteria for decision making in pediatric critical care units, which the "Baby Doe" regulations and subsequent Child Abuse regulations have expressly prohibited. Perhaps most important, it is not consistent with federal and state laws that tend to define citizens in this society as legally being persons from the time of their births.

Nevertheless, as graphically illustrated by the ongoing controversy in this society over abortion, questions about the ontological status of human beings in certain situations do not simply go away. In time, they occasionally influence changes in laws and public policy, as has already happened regarding abortion laws and may happen again.

I frankly doubt that public policy regarding severely disabled neonates will ever be based on the philosophical view that most neonates are regarded as potential persons. However, I remain convinced that the concepts of personhood and potential personhood have some merit in the prolonged, often agonizing deliberations that take place about neonates having severe to profound neurological impairments. At the very least, the concept of potential personhood provides an intellectual framework in which prognostic judgments about the future level of a baby's neurological development or future degree of a baby's neurological impairment make some sense. After all, at least part of the uncertainty that everyone struggles to get a handle on has to do with who a particular baby has the chance to become or, put more bluntly, whether the baby in question has the potential even to become a who in the normal course of his or her development.

What Is the Best Ethical Option for Making Decisions to Initiate, Continue, or Abate Life-Sustaining Treatment?

This question addresses the uncertainty generated by the pluralism of value perspectives over how life-and-death decisions should be made in pediatric critical care units and by whom. This pluralism is often present as neonatologists discuss treatment options with parents, pediatric specialists and residents discuss problematic cases during rounds, professionals from multiple health-care fields exchange views during case conferences, and pediatric ethics committees struggle to give appropriate advice regarding cases referred to them.

Each of the cases presented earlier illustrates the importance of this second question. In case 1, the pediatric specialists and the parents of baby girl A were obviously at odds over what the morally correct decision was, as well as who should have the moral and legal right to make the decision; the members of the ICRC were also far from agreement on these matters. The divergence of value perspectives could hardly be more vividly illustrated than it is in case 2, with Sammy Linares' father finally resorting to the use of a gun to carry out what he believed to be the morally correct decision. Case 3 shows an alternative way of resolving a fundamental difference in value perspective—a difference so great that the physicians denied a parental demand for life-sustaining treatment even in the face of judicial intervention in the case. Cases 4 and 5 demonstrate that medical uncertainty exacerbates the differing ethical perspectives among health-care professionals on problematic cases, that indecisiveness and vacillation on the part of parents also contribute to a divergence of opinion among health-care professionals regarding the appropriate course of action, and that this pluralism in ascertaining the morally appropriate (and legally permissible) course of action can result in enormous stress for physicians and nurses, and enormous expense for whoever (parents, hospital, third-party payer) ends up paying the bill.

In these cases and other cases with neonates or young children with life-threatening medical conditions, the adults involved in the cases have five ethical options available to them for deciding about life-sustaining treatment (Weir, 1984). The options range from very conservative to very liberal, and they differ from one another regarding the substantive and, to a lesser extent, the procedural aspects of making life-and-death decisions for nonautonomous young patients.

The most conservative of these ethical options is the ethical perspective that was enacted into public policy by the Reagan administration through the "Baby Doe" regulations and subsequent Child Abuse regulations. Incensed that one "Baby Doe" (the 1982 Bloomington, Indiana, case) had died who could have lived with surgical intervention and concerned that other disabled infants were unnecessarily being allowed to die in other hospitals, the leaders of the Reagan administration went to great lengths to advocate the ethical perspective held by Surgeon General C. Everett Koop, and many of the administration's prolife supporters.

This ethical position holds that there is one and only one acceptable moral reason for not sustaining an infant's life, namely, the medical futility in a very limited number of cases of trying to do so. According to this ethical perspective, decisions not to sustain a severely disabled infant's life are acceptable only when such an infant is irretrievably dying (or, for some persons holding this position, an infant whose condition is some form of permanent unconsciousness). Therefore, the only cases in which such decisions by physicians or parents are justifiable are those unusual cases in which there is actually no moral decision to make: God, nature, fate, the roll of the genetic "dice," or some force beyond our control prevents medical efforts at sustaining life from working.

The most liberal option is a position that carries significant weight in some philosophical circles, but not, as we have already discussed, among physicians and others who are more oriented toward a practical, empirically based view of reality. This position, in sharp contrast to the first position, is based on the ontological status of the young lives at risk in critical care units rather than on the severity of their medical conditions. Instead of calling for life-sustaining treatment to be administered to all neonates or young children who are not dying (or permanently unconscious), the philosophers holding this position (e.g., Michael Tooley, Mary Anne Warren, and Peter Singer) argue that physicians and parents are obligated only to provide life-sustaining treatment for neonates and young children who count as persons. The catch is, as we have seen, that according to this perspective no neonates meet the criteria for personhood, and no moral weight is placed on the potential they may have to become persons later in the course of their development. An unresolved problem for these philosophers—and one of the reasons that this position will never become public policy—is that of defining the "magic moment" beyond the neonatal period when young children do meet the criteria for personhood and are thus protected from having their lives arbitrarily terminated.

The third position is the first of three positions that reside closer to the middle of the philosophical spectrum than either of the views just discussed. The physicians, philosophers, and other individuals who hold this view do not believe that all nondying neonates should be given life-sustaining treatment, nor do they believe that the lives of neonates can be terminated morally on the basis of a definitional point about personhood. Rather, they are convinced that the most important aspect of decisions not to sustain some infants' lives is the procedural question of who should make these difficult decisions. The correct answer to that question, according to the advocates of this position, is that the appropriate decision maker is the parent or parents of the neonate or young child whose life is threatened by his or her medical condition, even though the current federal regulations do not permit this kind of parental discretion. Since the parents of a disabled infant are the ones who stand to gain or lose the most, depending on what happens to the infant, it is they—instead of the physicians, an ethics committee, or anybody else—who should have the right to make the life-or-death decision in all cases over which there is some disagreement about whether a disabled infant should continue to live with or die in the absence of life-sustaining treatment.

Advocates of a fourth ethical position are convinced that quality-of-life judgments are unavoidable in cases of severe neurological or physiological malformation, in spite of what the federal regulations say to the contrary. All of the responsible parties in cases of serious neonatal abnormalities are morally obligated—and should be legally permitted—to raise important questions about the most likely future ahead of these children if their lives are to be prolonged with medical treatment. Of fundamental importance in such cases is not only the question of whether a given child can be salvaged with the abnormalities he or she has, but also what kind of life he or she is most likely to have with those abnormalities. The most important abnormalities to consider are neurological in nature. If a neurological disorder is sufficiently serious that pediatric neurologists and neonatologists project a life with severe disabilities for the child, virtually all persons holding a quality-of-life position would find the abatement of life-sustaining treatment in such a case to be morally justifiable.

The fifth position is held by individuals who are convinced that life-sustaining treatment should be provided to normal and disabled neonates whenever such treatment is in their best interests, and that life-sustaining treatment should be abated in the care of severely premature or severely disabled neonates (and other young children) whenever such treatment is judged not to be in their best interests.

Persons holding this position tend to be in agreement with quality-of-life advocates whose projections of a given child's future focus *entirely* on that child's likely abilities and disabilities, not on the child's impact on anybody else or ability to attain somebody else's minimal standard of acceptability for personal human life. By contrast, persons having a best-interests position disagree with quality-of-life advocates who tend to compare mentally and physically abnormal children with normal children, emphasize the problems that disabled children cause for their families and society, and try to protect families and society from having to deal with disabled children who cannot meet some arbitrary standard of acceptability. Like all advocates of the quality-of-life position, proponents of the best-interests position hold a view that is more liberal than the current federal regulations.

The best-interests position, in my judgment, is the preferable ethical perspective to take in regard to difficult decisions about initiating, continuing, or abating life-sustaining treatment with any patients having life-threatening medical conditions. Neonatal and other young pediatric patients are no exception. They, like other nonautonomous patients, should receive life-sustaining treatment whenever the decision makers are convinced that the treatments available provide a balance of benefit to burden for the child. Such decisions should focus on the child's medical condition, con-

cern suffering and irremediable handicap rather than projected social worth, and involve comparative judgments about
the continuation of the child's injurious existence as opposed
to the child's nonexistence.

What Does "Best Interests" Mean
When Patients Are Neonates?

Even though widely supported in theory, the best-interests position is not without problems. Some of the advocates
of the position (President's Commission, 1983; McCormick,
1986; Jonsen et al., 1986; Bartholome, 1988) admit that the
concept of the patient's best interests is inherently vague,
especially when the patient is a neonate. Nevertheless, they
argue that the concept is helpful in decision making about
life-sustaining treatment for neonates, because it focuses the
decision-making process on precisely the human lives that
ought to be the primary focus of concern.

Some of the critics of the best-interests position, at least
as it applies to neonates, think that the conceptual foundation on which it stands is fundamentally flawed. Martin
Benjamin (1983) argues that neonates simply do not yet possess the cognitive awareness, much less the specific wants
and purposes, that are necessary for ascribing to them an
interest in continued life. Howard Brody (1988) is convinced
that any attempt to apply the concept of best interests to
infants is bound to fail, because the concept is either incoherent or inadequate as a guide for tough clinical decisions.
He argues that, even if infants can intelligibly be said to
have interests, such interests would be unknowable by adult
decision makers.

The third question of uncertainty thus has to do with
the intellectual integrity and plausibility of the concept of
the patient's best interests when that concept is used in the
context of NICUs. Is "the best interests of the infant" only a
symbolic slogan without substance? Is its only function that
of oversimplifying the complex decisions that have to be made
in NICUs?

I think not, and I will make three points briefly to try to demonstrate the importance and usefulness of the concept. To begin, it is necessary to say something about interests. One's "interests" consist of relationships, activities, and things in which one has a stake and on which one places value. To have interests (as opposed to sensations or instincts) normally requires as necessary conditions that one be conscious, aware of oneself, and able cognitively to have wants and purposes. In other words, to have interests normally requires that one be a person.

Yet, as Joel Feinberg (1984) points out in a discussion of fetal interests, it is plausible to ascribe *future* interests to a "prepersonal fetus." Even though a fetus "presumably has no actual interests," it can correctly be said to have future interests on the assumption that it will at some future point in its normal development (at birth or subsequent to birth) become a person and, thus, the possessor of actual interests. In a similar manner, the law recognizes that fetuses can have "contingent rights," such as the right to property, that will become actual rights the moment the fetus becomes a baby. Any contingent right of a fetus is instantly voided if the fetus dies before birth.

The same kind of reasoning about interests can be used in analyzing the interests that are ascribable to neonates, even by philosophers who claim that all neonates are nonpersons. For even if neonates as nonpersons cannot correctly be said to be the possessors of actual interests, they can be said to have future interests (assuming that they will at some future point become persons) that can be interfered with or damaged by decisions or actions by adults long before these developing human lives become persons. For example, a neonate with myelomeningocele could reasonably be said to have a future interest in physical mobility, but come to realize later in life that a decision by physicians or parents during the neonatal period not to have the lesion surgically corrected had preempted that future interest from being actualized.

An alternative conceptual framework for discussing the interests of neonates was presented earlier, namely the philosophical view that neonates without severe neurological impairments are to be regarded as potential persons. In this framework, an analysis of the interests of neonates does not involve the ascription of future interests to them because they are thought likely to become persons at some "magic moment" in the future, but ascribes future interests to them because they have the potential to become the possessors of interests through the normal course of their development.

The point is a fundamental one. Just as potentiality is an important aspect of the concept of personhood, so potentiality is an important feature of a philosophical understanding of interests (but not of legal rights). Interests change from time to time after one becomes a person, with some interests intensifying over time, others waning, and others appearing as though newly born. For that reason, a discussion of the future interests of any given neonate becomes problematic if one can only project the actual interests that child will have when he or she meets the criteria for personhood at some future point in time. By contrast, the principle of potentiality, as it applies both to the possession of personhood and the possession of interests, permits one reasonably to ascribe to any given neonate the most general and basic kinds of interests that most individuals tend to have as they develop from young children to older children and on through the various phases of personal life.

Second, even if it is not incoherent to discuss the general interests (in the sense of future interests) that neonates have, one must still make some sense out of the concept of "best interests." Simply put, the concept has to do with the fact that persons tend to have a multiplicity of interests that help to define who they are, and what they hold to be worthy of their expenditures of time, money, effort, and prolonged thought. Given this multiplicity of interests, some of the interests are necessarily regarded as being more important than others.

A common way of focusing on the interests that are most vital to a person, especially when that person is a patient in a hospital, is through the language of "best interests." Of course patients, at least those who are conscious and autonomous, do not lose their "regular" interests (e.g., in their families, personal goals, and work) when they are admitted to a hospital. However, their interests related to their own health tend to take on an immediacy and overall importance that transcend the range of other interests they have. In particular, if they have an injury or illness that is life-threatening in nature, their normal interests in the prolongation of their lives and the avoidance of suffering are greatly intensified, to the point of becoming the focal point of all their interests at that time. The concept of the patient's "best interests" captures this change in the ranking of personal interests, especially when the clinical circumstances concern decisions about administering or abating life-sustaining treatment.

When applied to neonates, the concept of "best interests" can obviously not refer to the specific wants and purposes any given neonate may have in continued life, much less to the specific wants and purposes that the neonate may have later in life. However, the concept of "best interests" can be used to capture the most fundamental future interest that persons have when they are patients, namely, an interest in not being harmed on balance during the course of medical treatment. For most patients in most clinical situations, this vital interest in not being harmed on balance means that they prefer continued life to death—unless intractable pain and other suffering have made continued life more harmful than the prospect of death. To ascribe this general and basic interest to neonates is to claim that all neonates lacking severe neurological impairment can reasonably be said to have this future interest in not being harmed, an interest that will become actualized as they become persons during the normal course of their development.

Third, the toughest aspect of using the concept of best interests in decision making in NICUs is determining the factors that should be considered in any given case. How can physicians and parents assess the beneficial and detrimental aspects of medical treatment in a case? How can they decide if life-sustaining treatment is in a neonate's best interests or is contrary to those interests?

My suggestion is to regard the patient's-best-interests standard as having eight variables (Weir and Bale, 1989). In neonatal (and other young pediatric) cases, the variables are as follows:

1. Severity of the patient's medical condition;
2. Availability of curative or corrective treatment;
3. Achievability of important medical goals;
4. Presence of serious neurological impairments;
5. Extent of the infant's suffering;
6. Multiplicity of other serious medical problems;
7. Life expectancy of the infant; and
8. Proportionality of treatment-related benefits and burdens to the infant.

The last of these variables is, in many respects, a summation of the preceding variables. For decision makers in such cases, a consideration of the benefits of the treatment (both short-term and long-term) to the patient is the "bottom line" for determining whether life-sustaining treatment or the abatement of life-sustaining treatment is in a particular neonate's best interests. In making this assessment, decision makers arrive at a subjective judgment that includes objective factors, but is not finally reducible to quantifiable information. For to decide in rare clinical situations that treatment is, on balance, harmful to the infant rather than beneficial is to make a moral judgment.

Should the Lives of Severely Disabled Neonates and Other Young Children Always Be Sustained with Nutrition and Hydration?

This question of uncertainty reflects the ongoing debate in this society over the morality and legality of abating technological feeding and hydration. The societal debate over technological feeding has been characterized by exchanges in the media, the professional literature, the courts, and the legislatures between the advocates of a mainstream position and the proponents of a large minority position on this issue.

The advocates of the mainstream position claim that technologically supplied nutrition and hydration may be abated on moral and legal grounds, in the same way that decisions are made about other life-sustaining technologies, whenever such treatment is contrary to the preferences of an autonomous patient or the best interests of a nonautonomous patient. The proponents of the minority view argue that the provision of nutrition and hydration, even if supplied by any of several kinds of feeding tubes, cannot correctly be regarded as a form of life-sustaining medical treatment or ever be discontinued in the care of nonautonomous patients (Weir, 1989).

In addition, this question of uncertainty reflects a fundamental ambiguity in the Child Abuse regulations (Department of Health and Human Services, 1985). Three exceptions to the requirement of providing medically indicated treatment are spelled out in the regulations:

1. Infants who are "chronically and irreversibly comatose;"
2. Dying infants for whom the provision of medical treatment would be "futile in terms of the survival of the infant;" and
3. Infants for whom the treatment would be "virtually futile . . . and . . . inhumane."

Each of these exceptions is described in ambiguous terminology, and each has been the subject of considerable discussion and debate since these exceptions were published in 1985.

An equally ambiguous part of the regulations immediately precedes the description of the three exceptions. The same sentence concerning medically indicated treatment states that effective treatment includes "appropriate nutrition, hydration, and medication." This statement, in contrast to the description of the three exceptional situations, has received relatively little attention in the literature on the regulations or in the more general literature on the issue of technological feeding.

For example, neither a questionnaire directed at Massachusetts pediatricians (Todres et al., 1988) nor a national survey of neonatologists (Kopelman et al., 1988) asked the physicians what they understood this portion of the regulations to mean. Likewise, the policy statements on technological feeding by medical and health-care organizations focus on adult cases, and do not even raise, with the exception of the statement of the American Nurses' Association (1988), the issue of technologically supplied nutrition and hydration for infants.

One exception to this trend is a short chapter written by Joel Frader (1986). Frader, a pediatrician at the University of Pittsburgh, argues that the same kind of reasoning about the provision of technological feeding "should apply equally well to infants and older persons, even though feeding babies has special social significance." He suggests that, in at least a few instances (e.g., hydranencephaly or renal agenesis), the provision of fluids and nutrition to neonates will only prolong their dying or extend their suffering. In these rare cases, he suggests that "good medical practice, that is, palliation, could require allowing dehydration and malnutrition."

Another exception is an article written by John Paris and Anne Fletcher (1987). Paris, an ethicist, and Fletcher, a pediatrician, emphasize that the language of "appropriate nutrition, hydration, and medication" allows for the kind of discretionary decision making that is required by complex NICU cases. They describe three clinical cases in which physicians and parents decided against the continuation of technological feedings: a premature baby with necrotizing enterocolitis, a neonate with severe birth asphyxia, and a three-month-old girl dying with multiple medical complications incompatible with long-term survival. In each of these cases, Paris and Fletcher think that the abatement of technologically supplied nutrition and hydration was medically and ethically appropriate.

These two publications notwithstanding, many neo-natologists and nurses in NICUs have a great deal of uncertainty about the morality and legality of abating life-sustaining nutrition and hydration. They remain uncertain even when informed of the mainstream ethical and legal position that says: (1) that the technological provision of nutri-tion and hydration is not morally or legally different from other life-sustaining care, and (2) that decisions about abating life-sustaining treatment with neonates and other young children should follow the best-interests standard in a manner similar to decisions about life-sustaining treat-ment made on the behalf of nonautonomous, adult patients. Their uncertainty is partially the result of the ambiguities they perceive in the law, and partially the result of the emo-tional and psychological realities of discontinuing nutrition and hydration in the care of babies with life-threatening conditions.

I anticipate that this uncertainty will be lessened as neo-natologists and nurses in NICUs do more to address the ques-tion of when technological nutrition and hydration are and are not appropriate. At the present time, however, their uncertainty about this question is very real.

Should Life-Sustaining Treatment for Neonates and Other Young Children Ever Be Abated for Economic Reasons?

Neonatologists and other pediatric specialists place con-siderable importance on providing good patient care. In terms of the patient's-best-interests position, this emphasis on the medical needs and interests of individual patients is the mor-ally preferable perspective for pediatricians and other physi-cians to have. According to this view, the needs and interests of each patient related to continued life correctly outweigh any competing interests of parents, siblings, or society. Sim-ply put, no neonate or other young, nonautonomous patient should die merely because their medical and hospital care is expensive, even when the physicians and parents in a given case know that the family's income and insurance cannot cover the costs involved in the patient's care.

In the last few years, however, a number of factors have combined to create uncertainty about this basic moral premise for the provision of medical care, especially as it applies to cases of extremely premature or severely disabled neonates. Physicians, hospital administrators, and other concerned persons often question the importance that should be placed on the economic aspects of sustaining the lives of some neonates and other young children, especially when these lives predictably will be characterized by severe mental and physical disabilities. Case discussions in NICUs, PICUs, and specialized chronic care units for young children increasingly have comments and questions by staff physicians, residents, nurses, social workers, and ethicists regarding the costs of the ongoing treatment and who will have to pay for those costs.

One of the factors contributing to this uncertainty about the economics of neonatal care has to do with the public policy regarding medically indicated treatment for disabled young children developed by the Reagan administration and continued by the Bush administration. Whether the legal framework for the policy was the 1973 Rehabilitation Act (for the now invalidated "Baby Doe" regulations) or the 1984 Child Abuse Amendments (for the current Child Abuse regulations), the public policy developed by the federal government has placed unprecedented pressure on neonatologists and parents to make treatment decisions without regard to cost.

The unfortunate and often tragic catch has been that the government, in mandating the administration of life-sustaining treatment for all neonates who are not dying or permanently unconscious, has never provided adequate financial coverage for that treatment for hospitals or for parents who are uninsured, underinsured, or have limited eligibility for state aid programs for disabled children. Moreover, given the drastic cuts by the Reagan administration in Medicaid funds and in funds for maternal and child health services, the various state aid programs for disabled children have much more restricted budgets now than they did in the early 1980s.

Another factor contributing to uncertainty about the role of economics in neonatal cases pertains to the escalating costs of the care needed by extremely premature or severely disabled newborns. This uncertainty is brought about not only by an awareness of the escalating costs of providing care for these babies, but also by the realization that efforts to provide comparative cost figures for neonatal care have proven less than satisfactory, that the application of diagnosis-related group (DRG) categories has not worked well in NICUs, and that the cost-effectiveness of neonatal care for low-birthweight neonates is still questionable (Berki and Schneier, 1987; Lichtig et al., 1989; Young and Stevenson, 1990).

The escalating cost of providing neonatal intensive care is widely recognized by neonatologists and hospital administrators, even though the medical diversity of NICU cases and variation of financial charges from hospital to hospital (and country to country) make accurate comparative cost comparisons virtually impossible to determine. At the very least, the figure of $8000 that a 1981 study by the Office of Technology Assessment gave as the average expenditure per NICU patient now seems quite low (Budetti, 1981).

Studies published in recent years all document the increasing cost of providing care for disabled neonates and young children in chronic care units. For example, one study from Canada (using 1978 Canadian dollars) found that the costs of intensive care for infants weighing less than 1000 grams averaged $102,500 per survivor (Boyle et al., 1983). A study from Australia (using 1984 Australian dollars) determined that the total direct cost for level-III, high-dependency care in one hospital was $690 per day (Marshall et al., 1989). Studies in the United States, varying greatly in methodology, have found the total cost for selected survivors of neonatal intensive care in a Boston hospital to range from $14,600 to $40,700 (Kaufman and Shepard, 1982), for long-term survivors in a Washington, DC, pediatric hospital to be $182,500 for a year (Battle, 1987), and for extremely low-birthweight survivors of NICUs in six medical centers to range from $72,110 to $524,110, with a mean cost of care per infant of $158,800 for 137 days (Hack and Fanaroff, 1986, 1989).

A third factor has been the increased recognition that the financial pressures created by expensive neonatal and pediatric treatment can greatly damage and sometimes destroy families (Strong, 1983; Lyon, 1985). For example, a 1988 Minnesota followup study of disabled infants and their families had a number of disturbing findings: the proportion of families with young children, but lacking health insurance is increasing, 16% of the families in the study pay the entire cost of their health insurance, middle-income families have not qualified for state financial assistance, several of the families have filed for bankruptcy, and at least one family still owes a hospital and physicians over $300,000 for the care of their young child. The report concludes: "Families should not have to lose their homes, mortgage their future, or neglect other children's needs to pay for the care of a chronically ill or disabled child" (Van Allen et al., 1988).

A related, but different factor has to do with the long-term costs of providing medical, nursing, and surgical care for severely disabled children who remain in hospitals for months and years. Sometimes called "boarder babies," these children have complicated, chronic medical conditions, are usually dependent on mechanical ventilation and other technological assistance for survival, and frequently come from low-income, single-parent families that simply cannot afford (in terms of money and time) to have the child at home.

If no other institutional home can be arranged (usually because of the cost and technology involved), and if foster parents are not a realistic option, such children may reside for several years in a specialized chronic care unit in the hospital in which they were born. When that happens, the children become living symbols of a "second generation" type of problem brought about by the successes of neonatal intensive care: They are survivors of the NICU, but remain captives of medical technology in an institution that nobody would choose to call home. As Case 5 illustrates, they also become the focal point for an ongoing debate among health professionals about the justifiability of the high costs required to keep them alive.

Given the uncertainty generated by these variables, what should be done? For two of the ethical positions described earlier, the answer is reasonably simple: revise or ignore the federal regulations, abate life-sustaining treatment more quickly on the basis of (1) parental discretion or (2) projected quality of life for the neonates involved (including the impact of a neonate's later life on others), and thus cut down on the costs in NICUs, to families, and to institutions.

To go that way, for that reason, would be a mistake. The economic aspects of neonatal intensive care would become a dominant factor in decision making by parents and physicians, and many premature and disabled neonates would have their lives cut short to save money. To establish a policy that would encourage parents to make life-and-death decisions in individual cases as a money-saving strategy for themselves (or for physicians to do the same to save money for their hospitals) is not the best policy for addressing the very real problem of escalating costs for neonatal intensive care, especially if that policy is to be guided by the ethical principles of beneficence, nonmaleficence, and justice.

There is, in my judgment, a better alternative. That alternative is a combination of:

1. Continued use of the patient's-best-interests standard in clinical settings, including increased emphasis on the eight variables that comprise the standard;
2. The establishment of a national policy, based on sound clinical evidence, that would restrict the use of neonatal intensive care in terms of infants' birthweights; and

3. The establishment of a national health insurance program that would pay for the catastrophic health-care expenses generated by providing care for extremely premature and severely disabled newborns.

The results of this combined approach would be three-fold. A more consistent application of the best-interests standard would result in an increased number of decisions, as difficult as they are, by parents and physicians to discontinue life-sustaining treatment in individual cases. Such decisions would not be made to save money, but would be based on an honest conclusion that the treatment available, although capable of sustaining a neonate's life, is contrary to the infant's best interests.

In addition, the establishment of a national policy that would limit life-sustaining treatment to neonates over a certain birthweight (e.g., 600 grams) would not only cut down on the enormously high costs of caring for extremely low-birthweight infants, but could also be defended, depending on the rationale and details of the policy, as meeting the requirements of justice. Such a policy would surely not solve all of the problems of uncertainty in NICUs, but could provide a measure of greater certainty, if based on a consensus among neonatologists, in establishing a minimum weight limit for neonates who would be given life-sustaining treatment.

Finally, by establishing a national insurance program, the federal government would help pay for the enormous costs that are involved in neonatal intensive care and specialized chronic-care units for young children. For the federal government to mandate that virtually all neonates, unless dying or permanently unconscious, be kept alive, and then to make no serious effort to help parents and institutions pay for that expensive care is unjust. In the absence of such a program, parents and physicians will continue to be faced with the task of making life-and-death decisions for newborns in the midst of great uncertainty—including whether the family will be destroyed financially by costs of the medical care.

References

American Nurses' Association (1988) Guidelines on withdrawing or withholding food and fluid.

Bartholome, W. G. (1988) Contested terrain: In the best interests of . . . *Hastings Cent. Rep.* **18,** 39, 40.

Battle, C. U. (1987) Obligations to the survivors of technology, in *Clinics in Perinatology*, vol. 14 (Silber, T. J., ed.), W. B. Saunders, Philadelphia, pp. 419–425.

Benjamin, M. (1983) The newborn's interest in continued life: A sentimental fiction. *Bioethics Rep.*. **1,** 5–7.

Berki, S. E. and Schneier, N. B. (1987) Frequency and cost of diagnosis-related group outliers among newborns. *Pediatrics* **79,** 874–881.

Boyle, M. H., Torrance, G. W., Sinclair, J. C., and Horwood, S. P. (1983) Economic evaluation of neonatal intensive care of very-low-birth-weight infants. *N. Engl. J. Med.* **308,** 1330–1337.

Brody, H. (1988) Contested terrain: In the best interests of . . . *Hastings Cent. Rep.* **18,** 37–39.

Budetti, P., et al. (1981) *Case Study #10: The Costs and Effectiveness of Neonatal Intensive Care* (Office of Technology Assessment, Washington, DC).

Department of Health and Human Services (1985) Child abuse and neglect prevention and treatment program. *Fed. Reg.* **50,** 14878–14901.

Engelhardt, H. T. Jr., (1986) *The Foundations of Bioethics* (Oxford University Press, New York), p. 117.

Feinberg, J. (1984) *Harm to Others* (Oxford Univerisity Press, New York), p. 96.

Feinberg, J. (1986) Abortion, in *Matters of Life and Death* (Regan, T., ed.), Random House, New York, pp. 256–292.

Frader, J. (1986) Forgoing life-sustaining food and water: Newborns, in *By No Extraordinary Means* (Lynn, J., ed.), Indiana University Press, Bloomington, IN, pp. 180–185.

Goldman, G. M., Stratton, K., and Brown, M. D. (1989) What actually happened: An informed review of the Linares incident. *Law, Medicine and Health Care* **17,** 298–307.

Gostin, L. (1989) Editor's introduction: Family privacy and persistent vegetative state. *Law, Medicine and Health Care* **17,** 295–297.

Hack, M. and Fanaroff, A. A. (1986) Changes in the delivery room care of the extremely small infant: Effects on morbidity and outcome. *N. Engl. J. Med.* **314,** 660–664.

Hack, M. and Fanaroff, A. A. (1989) Outcomes of extremely low-birth-weight infants between 1982 and 1988. *N. Engl. J. Med.* **321,** 1642–1647.

Jecker, N. S. (1990) Commentary: The moral status of patients who are not strict persons. *J. Clin. Ethics* **1,** 35–38.

Jonsen, A., Siegler, M., and Winslade, W. J. (1986) *Clinical Ethics* (Macmillan, New York), p. 188.

Kaufman, S. L. and Shepard, D. (1982) Costs of neonatal intensive care by day of stay. *Inquiry* **19,** 167–178.

Kopelman, L. M., Irons, T., and Kopelman, A. (1988) Neonatologists judge the "Baby Doe" regulations. *N. Engl. J. Med.* **318,** 677–683.

Lichtig, L. K., Knauf, R. A., Bartoletti, A., Wozniak, L. M, Gregg, R. H., Muldoon, J., and Ellis, W. C. (1989) Revising diagnosis-related groups for neonates. *Pediatrics* **84,** 49–61.

Lyon, J. (1985) *Playing God in the Nursery* (W. W. Norton, New York), p. 237.

McCormick, R. A. (1986) The best interests of the baby. *Second Opinion* **2,** 18–25.

Marshall, P. B., Halls, H. J., James, S. L., Grivell, A. R., Goldstein, A., and Berry, M. N. (1989) The cost of intensive and special care of the newborn. *Med. J. Australia* **150,** 568–573.

Murray, T. H. (1985) Why solutions continue to elude us. *Soc. Sci. Med.* **20,** 1102–1107.

Paris, J. J. and Fletcher, A. B. (1987) Withholding of nutrition and fluids in the hopelessly ill patient, in *Clinics in Perinatology,* vol. 14 (Silber, T. J., ed.), W. B. Saunders, Philadelphia, pp. 367–377.

Paris, J. J., Crone, R. K. and Reardon, F. (1990) Physicians' refusal of requested treatment: The case of Baby L. *N. Engl. J. Med.* **322,** 1012–1015.

Perkins, R. L. (1974) *Abortion Pro and Con* (Schenkman, Cambridge, MA), p. 229.

President's Commission for the Study of Ethical Problems in Medicine and Biomedical and Behavioral Research (1983) *Deciding to Forego Life-Sustaining Treatment* (US Government Printing Office, Washington, DC), p. 217.

Singer, P. (1979) *Practical Ethics* (Cambridge University Press, Cambridge, UK), p. 97.

Strong, C. (1983) Defective infants and their impact on families: Ethical and legal considerations. *Law, Medicine and Health Care* **11,** 168–181.

Todres, I. D., Guillemin, J., Grodin, M.A., and Batten, D. (1988) Life-saving therapy for newborns: A questionnaire survey in the state of Massachusetts. *Pediatrics* **81,** 643–649.

Tooley, M. (1972) Abortion and infanticide. *Philosophy and Public Affairs* **2,** 37–65.

Tooley, M. (1979) Decisions to terminate life and the concept of person, in *Ethical Issues Relating to Life and Death* (Ladd, J., ed.), Oxford University Press, New York, pp. 62–93.

Tooley, M. (1983) *Abortion and Infanticide* (Oxford University Press, New York), p. 332.

Van Allen, E., Johnson, D. E., McWilliams, D., and Weckwerth, V. (1988) *Life After the NICU: Meeting the Needs of Sick and Disabled Infants and Their Families* (Minnesota Hospital Association, Minneapolis).

Warren, M. A. (1973) On the moral and legal status of abortion. *The Monist* **57,** 43–61.

Weir, R. F. (1989) *Abating Treatment with Critically Ill Patients* (Oxford University Press, New York).

Weir, R. F. (1986) When is it justifiable not to treat? *Second Opinion.* **2,** 42–61.

Weir, R. F. (1985) Selective nontreatment—one year later: Reflections and response. *Soc. Sci. Med.* **20,** 1109–1117.

Weir, R. F. (1984) *Selective Nontreatment of Handicapped Newborns* (Oxford University Press, New York), p. 143.

Weir, R. F. and Bale, J. F., Jr. (1989) Selective nontreatment of neurologically impaired neonates, in *Neurologic Clinics*, vol. 7 (Bernat, J. L., ed.), W. B. Saunders, Philadelphia, pp. 807–822.

Young, E. W. D. and Stevenson, D. K. (1990) Limiting treatment for extremely premature, low-birth-weight infants (500 to 750 g). *Am. J. Dis. Child.* **144,** 549–552.

Conflict, Compromise, and Symbolism

The Politics of the Baby Doe Debate

Janna C. Merrick

In September 1989, the US Civil Rights Commission released its report on medical discrimination against children with disabilities,[1] finding that its "inquiry leaves no doubt that newborn children have been denied food, water, and medical treatment solely because they are, or are perceived to be, disabled" (Civil Rights, 1989).

The Commission's chairman, William B. Allen, disavowed the report, arguing: "I remain convinced that this report reflects all too prominently a certain kind of research incontinence. . . . I [am] persuaded . . . that the interests of handicapped newborns have been sacrificed to a political mission" (Civil Rights, 1989).

The release of the report, accompanied by nationwide news coverage, set the stage once again for public debate concerning government's role in deciding treatment for newborns with disabilities. Traditionally, parents and physicians have acted as primary decision makers in choosing medical care for minor children. This pattern was challenged in the early 1980s as the result of a series of well-publicized

From: *Compelled Compassion* Eds.: Caplan, Blank, and Merrick
©1992 The Humana Press Inc.

events involving treatment practices for infants who were born with life-threatening defects and who were left untreated.

One specific case generated immense national publicity and became a catalyst for government regulation. In April 1982, Baby Doe was born in Bloomington, Indiana, suffering from Down syndrome and an abnormal, but surgically correctable, esophagus. Without surgery, the infant would die. The parents refused consent not only for the surgery, but also for intravenous feeding, which would have provided nutrition and hydration. Bloomington Hospital petitioned the Superior Court of Monroe County to determine the legality of the parents' decision. The court upheld them and ordered the hospital to comply with nontreatment (*Doe v. Bloomington Hosp.*, 1983). The Indiana Supreme Court refused to hear the case, and the baby died before an appeal could be made to the US Supreme Court. Thus began the public debate over the treatment of newborns with disabilities, and over who should have the right and responsibility for making these life-and-death decisions. The debate was thrust onto the public agenda through the intensive efforts of the prolife lobby, which actively pursued allies among disabled citizens rights groups and the medical profession.

In light of the Bloomington case—and pressure from the prolife lobby—the Reagan administration notified hospitals that failure to treat such infants was a violation of their civil rights. This approach was struck down by the US Supreme Court. Congress also became involved, enacting legislation in 1984 that made medical neglect of newborns with disabilities a form of child abuse.

The Policy Process

A number of books have been written on the plight of newborns with disabilities. The focus of this book is the *public policy* aspects of the Baby Doe debate. A useful place to begin, therefore, is with an analysis of the context in which public policy is made. The American political system is characterized by substantial government fragmentation. The Constitution establishes a separation-of-powers principle whereby policy-making functions are shared by the president, Congress, and the courts. Fragmentation is further exacerbated because the US is a federal system in which power is shared between the federal government and the 50 state governments. Thus, the federal government often relies on states for implementation of national policies.

Politics in the US is also characterized by conflict among adversaries, which is normally resolved through bargaining and compromise. Additionally, much of what goes on in the political system is more symbolic than substantive. Murray Edelman argues that the most cherished forms of popular participation in government are largely symbolic, and that politics for most of us is a passing parade of abstract symbols. He distinguishes between referential symbols, which depict concrete situations, and condensational symbols, which evoke emotions.

> Practically every political act that is controversial or regarded as really important is bound to serve in part as a condensation [*sic*] symbol. It evokes a quiescent or an aroused mass response because it symbolizes a threat or reassurance. Because the meaning of the act in these cases depends only partly or not at all upon its objective consequences, which the mass public cannot know, the meaning can only come from the psychological needs of the respondents; and it can only be known from their responses (Edelman, 1964).

Moreover, according to Edelman, a large number of people see and think in terms of stereotypes and oversimplifications. They cannot recognize or tolerate complex situations, and respond chiefly to symbols that distort (Edelman, 1964).

Policy making is a complex and usually lengthy process, and according to political scientist James E. Anderson involves five stages (Anderson, 1990).

1. *Problem identification and agenda formation.* The American nation is beset by a multitude of problems, of which only a few receive public recognition and fewer yet result in government action. Those issues that are currently matters of public debate are considered to be agenda items. In reality, according to Roger W. Cobb and Charles D. Elder, two agendas are formed:

> The systemic agenda consists of all issues that are commonly perceived by members of the political community as meriting public attention and as involving matters within the legitimate jurisdiction of existing governmental authority (Cobb and Elder, 1972).

These are issues that have come to the public's attention, and that the public is concerned about. A number of factors help push an issue onto the systemic—or public—agenda. Government officials, especially the president, play a major role, as do the news media. Cobb and Elder argue that unforeseen "triggering" events, including unanticipated human events (such as the death of Baby Doe) and technological change (such as the development of sophisticated neonatal life support systems), also push an issue onto the public agenda (Cobb and Elder, 1972).

A second agenda is the formal—or governmental—agenda, which includes a "set of items explicitly up for active and serious consideration of authoritative decision-makers" (Cobb and Elder, 1972). Examples include lawsuits in the courts, legislative proposals in Congress, and rules or regulations promulgated by administrative agencies. An issue reaches the formal agenda because policy makers feel compelled to act, or at least feel compelled to appear to be acting, to resolve the problem (Anderson, 1990).

2. *Formulation.* This is a complex process involving a wide range of actors inside and outside of government. In the case of Baby Doe policy, interest groups representing the prolife movement, disability rights organizations, and the medical profession were actively involved. All three

branches of the federal government participated, including officials of the Reagan administration, members of Congress, and the courts.

3. *Adoption.* Policy adoption typically takes the form of legislative enactment or presidential directive, with the possibility of court interpretation. The final product in the Baby Doe debate was legislation making failure to treat newborns with disabilities a form of child abuse, and requiring states receiving federal funds for child abuse programs to establish programs and procedures to respond to reports of medical neglect.

4. *Implementation.* Once enacted, the policy process shifts to the administrative branch for implementation. Agencies use a variety of tools including rule making, adjudication, law enforcement, and program operations. Rules explain how the legislation is to be applied, adjudication is the actual application of the rules to specific situations, law enforcement means that the agency uses its discretion to enforce the rules, and program operations involve other management aspects of the program (Anderson, 1990). Often policy formulation, adoption, and implementation are closely intertwined, since the policy process is generally one of continual incremental change.

5. *Evaluation.* Once policy is enacted and implemented, it is important to evaluate its impact. What *change*, if any, has taken place, and has that change been positive or negative? Anderson warns evaluators to distinguish between policy outputs and policy outcomes. Outputs "are the things actually done by agencies in pursuance of policy decisions and statements. The concept of outputs focuses one's attention on such matters as amounts of taxes collected, miles of highways built" (Anderson, 1990), and perhaps in the cases of Baby Doe, numbers of child abuse reports made. Outcomes "are the consequences, for society, intended and unintended, that stem from governmental action or inaction" (Anderson, 1990). The intended outcome of the Child Abuse Amendments was to ensure that newborns receive medically beneficial treatment regardless of handicaps.

Although this seems on the surface to be a simple and straightforward approach, in reality policy making in the US is extremely complicated, particularly in areas such as medicine, in which biomedical technology may have outdistanced reasoned ethical and policy analysis. This chapter will begin by discussing how the issue of treating newborns with disabilities evolved, and how the Department of Health and Human Services (DHHS) and the federal courts became entangled in a legal struggle. It will proceed to analyze the role of Congress and conclude with an evaluation of neonatal care in the post-Baby Doe period.

Building the Policy Agenda

Nationwide attention focused on the Bloomington case, which was clearly the most significant factor in arousing the public's concern for the treatment of newborns with disabilities, but this was only one in a series of events that came together to form the policy agenda. In this regard, Barbara Nelson's analysis of agenda setting is helpful. In discussing the systemic agenda, she makes an important distinction between the professional agenda, which reflects the views of those who are informed about a given issue and who may promote a particular point of view, and the popular agenda, which reflects the concerns of the mass public (Nelson, 1984). In the case of newborns with disabilities, both types of agendas played important roles.

For the most part, formation of the professional agenda preceded the popular agenda. In 1973, Raymond S. Duff and A. G. M. Campbell published their now-famous study of the special care nursery at the Yale–New Haven Hospital in the *New England Journal of Medicine* (Duff and Campbell, 1973). During a 30-month period beginning in January 1970, 43 deaths (14% of total deaths) in the nursery were related to withholding treatment. Duff and Campbell discussed extensively the complexity of decision making in the special care nursery, and the moral dilemmas facing staff and parents. They argued that, as policy:

> Since families primarily must live with and are most affected by the decisions, it therefore appears that society and the health professions should provide only general guidelines for decision making. Moreover, since variations between situations are so great, and the situations themselves so complex, it follows that much latitude in decision making should be expected and tolerated (Duff and Campbell, 1973).

They concluded that, if this policy, or their treatment decisions based on it, violated the law, then the law ought to be changed.

There were other studies as well. A survey of San Francisco area obstetricians and pediatricians found that 22% of respondents favored active or passive euthanasia for Down syndrome infants with no complications, and 50% favored euthanasia if the Down syndrome infant had an intestinal blockage (Treating the Defective Newborn, 1976). A survey of Massachusetts pediatricians showed that a significant majority (79.6%) believed that informed consent should include the right of parents to withhold consent for surgery and that, in a case of Down syndrome with a duodenal atresia and parental refusal for surgery, only 46.3% would recommend surgery. Of those recommending surgery in such cases, less than half (40.2%) would pursue a court order (Todres et al., 1977).

A national study in 1977 headed by Anthony Shaw and his colleagues showed that 50% of pediatricians and 77% of pediatric surgeons would not operate on a Down syndrome infant with an intestinal atresia if that decision was the parents' choice (Shaw et al., 1977). In the same year, Shaw published a formula in which he urged physicians and others to consider the contributions of the family and society, as well as physical and intellectual endowments, in determining quality of life. He clearly did not, however, advocate using the formula as a method of calculating the value of human life, defining humanhood, or assigning points to determine when to withhold treatment (Shaw, 1977).

In 1983, physicians at the University of Oklahoma Health Sciences Center published the results of their treatment methods during a five-year period beginning in July 1977, for infants with myelomeningocele. They indicated they had been "influenced" by Shaw's formula and stated that "treatment for babies with identical 'selection criteria' could be quite different, depending on the contribution from home and society" (Gross et al., 1983). According to Norman Fost, chair of the American Academy of Pediatrics (AAP) Bioethics Committee, this rate of nontreatment was much higher than had been reported elsewhere in the US, and thus, the research caused extensive controversy (Fost, 1986).

As the professional agenda became widely known, it generated a public reaction that became the popular agenda. Concern in the general public had begun to develop earlier because of the well-publicized death of a Down syndrome child with an intestinal blockage at Johns Hopkins University in 1971. Doctors had urged treatment, but the parents refused, and in fact, orders not to feed the child were taped to his bassinet as he lay dying for 15 days in an isolated section of the infant floor. Public interest in the case increased when the Joseph P. Kennedy Foundation filmed a documentary about it.

In 1981, Siamese twins were born in critical condition in Danville, Illinois. Neonatologists agreed that the twins would not survive (Staver, 1981). The facts and events are not clear, but it is believed that the parents—he a physician and she a nurse—requested that the children not be resuscitated. When they began to breathe spontaneously, a note stating that they were not to be fed in accordance with the parents' wishes was written on their chart.[2] After an anonymous telephone call, Illinois Department of Children and Family Services filed a neglect petition, and subsequently, the parents and attending physician were charged with conspiracy to commit murder. The charges were later dropped owing to lack of evidence (Weir, 1984), but media coverage was extensive.

The most significant and culminating event in setting the popular agenda was the slow death in 1982 of Baby Doe in Bloomington. It became a media event as attorneys and county prosecutors sought to shift custody away from the parents in order to force medical treatment. Other couples went on television offering to adopt the child. The Neonatal Intensive Care Unit (NICU) nurses refused to participate in his care—or lack thereof. Hours before arguments were to be heard by US Supreme Court Justice John Paul Stevens, Baby Doe died, and a national furor began. Prolife and disability rights groups—among others—expressed outrage. Columnist George Will, a friend of President Reagan and father of a Down syndrome child, editorialized that it was murder, writing that "the baby was killed because it was retarded . . . such homicides in hospitals are common and will become more so now that a state's courts have given them an imprimatur" (Will, 1982).

This was followed by an investigative report aired by the Boston CBS affiliate in the spring of 1983. "Death in the Nursery" covered the events in Bloomington, and alleged that infanticide was a hidden, but widespread, phenomenon in the nation's hospital nurseries. It also included interviews with two elementary school children who had been treated as infants by Duff and whose parents claimed he had advocated withdrawing treatment from them. The mood of the report was accusatory, and the public response was swift and sympathetic. Its role as a condensational symbol was powerful. News coverage continued with the birth of Baby Jane Doe in Port Jefferson, New York, in October 1983. She suffered from spina bifida and related complications. Her parents refused consent to surgery, and under doctor's advice pursued a "conservative" treatment involving antibiotics and dressings. Media coverage was extensive as the parents, hospital, DHHS, and a prolife attorney battled over access to her medical records and control of treatment decisions.

News media attention continued to focus on decision making in NICUs when another investigative report, entitled "Who Lives, Who Dies?", aired on Cable News Network in February 1984. It focused on the selection criteria reported by Gross in Oklahoma. The "camera captured the picture of Carlton Johnson, a black baby born with spina bifida to an unmarried mother receiving welfare" (Civil Rights, 1989). The report indicated that the child had not been given surgery and that the mother had been told her child could not live (Cable News Network Transcript, "Who Lives, Who Dies?" 3 [Feb. 21, 22, and 24, 1984], cited in Civil Rights, 1989). Some alleged that this type of selection process constituted racial and socioeconomic discrimination. Disability rights organizations filed formal complaints with DHHS and the Department of Justice in February 1984, and the American Civil Liberties Union, in conjunction with the National Center for the Medically Dependent and Disabled, filed suit against the physicians in October 1985 (Civil Rights, 1989).

Policy Formulation, Adoption, and Implementation Stage I: Withholding Treatment as a Form of Discrimination

When President Reagan learned of the Bloomington case, he ordered DHHS to issue a letter to hospitals receiving federal aid. It was based on section 504 of the 1973 Rehabilitation Act, which provides that:

> No otherwise qualified handicapped individual in the United States . . . shall solely by reason of his handicap, be excluded from the participation in, be denied the benefits of, or be subjected to discrimination under any program or activity receiving federal financial assistance (Rehabilitation Act of 1973, 1982).

DHHS notified health-care providers that it was unlawful for hospitals receiving federal aid to withhold treatment based on an infant's handicaps (Notice to Health Care Providers, 1982). This was an interesting approach considering that, one month earlier, the Reagan administration had proposed a 30% cut in funds for handicapped education and was in the process of drafting guidelines to limit federal enforcement of section 504 for adults by removing requirements that employers provide "equal opportunity" to the handicapped (*Wa. Post*, Mar. 4, 1982).

Prolife groups had already begun to mobilize their efforts. They were angry about the Reagan administration's seeming failure to fight more aggressively against abortion, and they wanted to improve their public image. In Edelman's terms, they would use evocative condensational symbols to help in their cause to gain public sympathy by arguing that newborns with disabilities were being mistreated and allowed to die on a regular basis in America's hospitals. The prolife movement considered human life to be sacred at any point and believed that securing protection for disabled newborns would give added basis for protecting the unborn fetus (Paige and Karnofsky, 1986).

Prolife groups began forging alliances. They turned to disability rights groups, which had become politically active in the 1970s and had seen some success in gaining legal protections for the disabled. These groups now wanted to expand their rights. This alliance proved difficult for two reasons: Disability rights groups were divided among themselves on strategy and goals, and most had supported prochoice positions on abortion.

Nevertheless, meetings began in late 1982 that included representatives from prolife and disability rights groups, and the Reagan administration. They developed a two-pronged strategy. First, they would search for actual cases in which newborns with disabilities were improperly treated, publicize them in the media, and begin litigation to force treatment. Second, they would lobby Congress in an effort to get legislation enacted. Prolife groups already had a network of friends both in the White House and on Capitol Hill (Paige and Karnofsky, 1986).

Prolife groups visited with the president and showed him the film *Death in the Nursery*. It is reported that he was deeply moved, arranged to have it shown on closed-circuit television in the White House, and encouraged staff to watch it. Members of the administration came to believe that newborns with disabilities were being denied treatment on a frequent basis, and it was thought that the president wanted quick action.

In March 1983, DHHS issued its first actual Baby Doe regulation. This "interim final rule" was drafted in a hurried fashion without the normal waiting period for public comment. It advised hospitals of the discrimination clause in the 1973 Rehabilitation Act, and required the posting of a notice in each maternity and pediatric ward, delivery room, and nursery. The notice stated that any person having knowledge that a handicapped infant was being discriminatorily denied food or customary medical care should immediately contact the "Handicapped Infant Hotline," and a 24-hour, toll-free number was provided. The notice also indicated that callers could contact their state child protective services (CPS) agency and that their identities would be kept confidential (Nondiscrimination on the Basis of Handicap, Interim Final Rule, 1983).

Procedures for implementing the rule provided that, upon receipt of calls, DHHS would contact the state CPS agency, which would follow its customary procedures in investigating allegations of child abuse and neglect. If direct federal intervention seemed appropriate, DHHS could conduct immediate on-site investigations, have access to medical records and facilities at any time, and make referrals to the Department of Justice for legal action.

The medical community objected to the rule, especially to the requirement for posting notices, which it felt implied that hospitals and staff discriminated against patients with disabilities. The AAP was vocal in its opposition, arguing that the interim final rule violated patient confidentiality as well as physicians' and hospitals' abilities to exercise professional medical judgment. It objected strongly to the use of a hotline. Moreover, it argued that the rule was excessively vague and simplistic, was disruptive to hospital–patient relationships, and had been promulgated in a hurried fashion without appropriate investigation on the part of DHHS. The AAP argued that DHHS had "not produced any direct evidence of epidemic, inappropriate treatment of severely ill newborns. Indeed, retrospective compliance reviews have apparently found all facilities to be in compliance with section 504" (Senate Hearings, April 6, 11, 14, 1983). The prestigious President's Commission on the Ethical and Legal Problems in Medicine and Biomedical and Behavioral Research (hereinafter referred to as the President's Commission) also criticized the adversarial atmosphere generated by the rule (President's Commission, 1983).

The AAP, the National Association of Children's Hospitals and Related Institutions, and the Children's Hospital National Medical Center filed suit in federal district court challenging the validity of the rule. In *American Academy of Pediatrics v. Heckler* (1983), the court invalidated the DHHS rule on procedural grounds, because the secretary failed to provide public notice or to grant a 30-day delay for its effective date. The court noted several additional issues. First, the rule did not consider the disruptive effects of a 24-hour, toll-free hotline. The court found that:

> Any anonymous tipster, for whatever personal motive, can trigger an investigation involving immediate inspection of hospital records and facilities and interviewing of involved families and medical personnel. In a desperate situation where medical decisions must be made on short notice by physicians, hospital personnel, and often distraught parents, the sudden descent of "Baby Doe" squads on the scene, monopolizing physician and nurse time and making hospital charts and records unavailable during treatment can hardly be presumed to produce higher quality care for the infant. (*AAP v. Heckler*, 1983).

The court also noted that the rule attempted to prevent parents from having influence on decisions when they were often in the best position to evaluate their child's interests. Finally, the court held that the rule was made in haste and ignorance, without considering the legal and constitutional issues that should have guided DHHS, and there was no indication that Congress ever intended for section 504 to apply to medical treatment for handicapped newborns.

In the spring of 1983, a Reagan administration task force began meeting with representatives from the AAP, the American Hospital Association, the American College of Obstetricians and Gynecologists, and the Children's Hospital National Medical Center. These groups argued that the best approach to dealing with treatment decisions was to establish individual hospital review committees to advise parents and doctors. The President's Commission had made similar recommendations. The task force was also meeting with representatives of prolife and disability rights organizations who claimed that retarded, but otherwise healthy, infants were being killed in the nation's hospitals because of the perception that their lives had less value than the lives of others. They were adamant that both the hotline and the threat of withdrawing Medicare and Medicaid funds from hospitals in noncompliance were essential (*Wa. Post,* May 23, 1983, March 30, 1983).

At this point, Surgeon General C. Everett Koop, a longtime supporter of the antiabortion movement and founder of the evangelical Christian Action Council, contacted the Secretary of DHHS indicating his desire to be involved in drafting the final rule. In a number of speeches and written statements, he emphasized his belief that infanticide was common in the nation's hospitals, and along with abortion and euthanasia, was a step along the "slide to Auschwitz" (Brown, 1986). This is the kind of language Edelman would describe as evocative, but not an instrument of reasoning or analysis. Considering Koop's standing as an eminent pediatric surgeon and as the surgeon general, it can be viewed as a very powerful use of condensational symbols. Despite his inflammatory public statements, Koop wanted to see a rule that prolife factions, disability rights groups, and physicians would accept, and he began negotiating compromise.

In July 1983, DHHS redrafted the rule to correct the procedural problems and invited comments. It received nearly 17,000 comments, 97% of which supported the rule. It is reported that the wording of most comments indicated they had been generated in response to requests from prolife organizations (Paige and Karnofsky, 1986). Language of the responses was symbolic and emotional. In Edelman's terms, it was the stale repetition of clichés, serving to evoke a conditioned, uncritical response and preventing systematic analysis of the situation (Edelman, 1964). Babies were being "murdered." Medical staffs had become "hardened to Death." Babies were dying through "malign neglect." Infant Care Review Committees were to handicapped infants as "foxes are to chickens" (Brown, 1986). Although the general response to the rule was favorable, the reaction from the medical community was mixed. Seventy-two percent of responding pediatricians opposed it, as did 77% of hospital officials and health-related associations. On the other hand, 98% of nurses responding supported the rule (Nondiscrimination on the Basis of Handicap, Final Rule, 1984).

Koop continued to urge compromise. In October 1983, the AAP published a statement in support of principles against withholding treatment, but at the same time objected to the mechanism by which the rule would be enforced. It concluded:

> The decision to treat or not treat seriously ill newborns should not be based on concomitant handicapping conditions, such as Down syndrome or other congenital anomalies . . . withholding or withdrawing life-sustaining treatment is justified only if such a course serves the interests of the patient. When the infant's prospects are for a life dominated by suffering, the concerns of the family may play a larger role.

Treatment should not be withheld for the primary
purpose of improving the psychological or social well
being of others, no matter how poignant those needs
may be. (AAP, 1983)

The AAP then joined with a number of medical and disability rights groups under the auspices of the Office for Special Education and Rehabilitative Services in the Department
of Education (Civil Rights, 1989) in issuing a joint statement
in November 1983.[3] It read in part:

Discrimination of any type against any individual with
a disability/disabilities, regardless of the nature or
severity of the disability, is morally and legally inde-
fensible. . . . Consideration such as anticipated or
actual limited potential of an individual and present
or future lack of available community resources are
irrelevant and must not determine the decision concerning medical care . . . the individual's medical condition should be the sole focus of the decision. (Joint
Policy Statement, 1984).

After considering these statements along with the
comments discussed above, DHHS promulgated the "final
rule" in January 1984, also based on section 504. It encouraged—but did not mandate—the establishment of Infant
Care Review Committees (ICRCs), which the President's
Commission, the AAP, and other medical groups had
strongly supported. It is reported that the inclusion of the
recommendation was a compromise solution urged by the
AAP, which then agreed not to challenge the rule (Civil
Rights, 1989). ICRCs would recommend institutional policies,
provide advice in specific cases, and retrospectively review
infant medical records. Membership would include physicians,
nurses, hospital administrators, representatives from both
the legal profession and disability rights groups, and lay
members (Nondiscrimination on the Basis of Handicap,
Final Rule, 1984). In making decisions, the committees would

rely on the guidelines offered in the "final rule" and would not be allowed to substitute their own interpretations. DHHS agreed that it would consult with the ICRC when conducting investigations, unless other action was necessary to protect the infant. It also encouraged state CPS agencies to do likewise (Civil Rights, 1989).

Notices indicating that denial of food or customary medical care was a violation of federal law would continue to be posted, but only where medical staff—rather than parents—would see them. They would include the DHHS 24-hour, toll-free "hotline" number, as well as the state CPS number. Hospitals would have the option of including information about their own treatment and internal review policies.

State CPS agencies receiving federal funds were required to establish and maintain procedures to prevent instances of medical neglect. They would notify DHHS of each reported instance and indicate what action had been taken. The rule also encouraged CPS agencies to work closely with ICRCs. Koop argued that state CPS agencies would be the "first line of defense" in enforcing the rule, and if those efforts failed, DHHS could become involved (*Washington Post*, Jan. 10, 1984). As under the "interim final rule," DHHS could conduct on-site investigations, and have 24-hour access to medical records and hospital facilities.

A significant change from the "interim final rule" was the inclusion of a section on interpretative guidelines dealing primarily with medically beneficial treatment and parental decisions.

> With respect to programs and activities receiving Federal Financial assistance, health care providers may not, solely on the basis of present or anticipated physical or mental impairments of an infant, withhold treatment or nourishment from the infant, who in spite of such impairments, will medically benefit from the treatment or nourishment (Nondiscrimination on the Basis of Handicap, Final Rule, 1984).

Physicians would not have to provide futile treatments that would simply prolong the process of dying, and could use "reasonable medical judgment" in selecting alternative courses. If parents refused treatment that was medically indicated, hospitals had an obligation to report them to the appropriate state authorities or seek court action. Quality-of-life considerations were clearly prohibited.

The AAP and the Association for Retarded Citizens supported the final rule with reservations. The American Life Lobby was hostile, describing it as a "cave-in to the medical/hospital industry" (*Washington Post*, Jan. 10, 1984) and arguing that the Reagan administration was "kowtowing to the medical establishment's God complex . . . accepting on even partial basis an institutional review board . . . is a venal decision and for handicapped babies, a fatal one" (*Washington Post*, Jan. 6, 1984).

Between the time that the "interim final rule" was overturned by the court and the "final rule" was promulgated, Baby Jane Doe was born and admitted to University Hospital at Stony Brook, New York. Litigation began in the New York State courts when a prolife lawyer who had worked for a New York branch of the National Right to Life Committee filed suit in Suffolk County Supreme Court seeking to force surgery. He coordinated his actions with the American Life Lobby, which financed much of the litigation. Baby Jane Doe's medical problems were complex, and this attorney had no access to her records and therefore no in-depth knowledge of her condition. Pursuing this case would, however, focus media attention on the plight of newborns with disabilities. Using Edelman's analysis, her condition was portrayed in oversimplified terms by the prolife movement, so that condensational symbols could be used to generate public sympathy and concern for the cause.

The state trial court authorized the surgery, but its appellate division reversed the decision, arguing that the parents had made reasonable choices among the various medical options and had the best interests of their child in mind (*Weber v. Stony Brook Hosp.*, 1983). The New York Court of

Appeals subsequently found that the trial court had erred in allowing a party with no relationship to the child, parents, family, or medical staff to bring suit because such action failed to comply with state law providing that child neglect proceedings be originated by child protective agencies or by a person on the court's direction (*Weber v. Stony Brook Hosp. Ct. App.*, 1983). During this time, DHHS received an anonymous complaint and referred it to New York's CPS agency, which conducted an investigation and concluded there was no cause for state intervention (Annas, 1986). Additionally, the American Life Lobby sent a telegram to DHHS about the case, and initiated a phone-calling and letter-writing campaign to it, the White House, and the Justice Department (Paige and Karnofsky, 1986).

Federal litigation involving Baby Jane Doe provided a constitutional challenge to the claim that Section 504 could be used to intervene in treatment decisions for handicapped newborns. It is reported that Surgeon General Koop personally telephoned Baby Jane Doe's surgeon, who refused to discuss the case. DHHS then sought her medical records and filed suit in federal court when the hospital refused to release them (*United States v. University Hosp.*, 1983). The federal district court found for the hospital, since it had failed to perform surgery because the parents had refused to give consent, and further found that the parents' choice was reasonable based on alternative treatment methods. Moreover, the hospital had refused to release the records because the parents had refused to authorize such a release. DHHS appealed to the Second Circuit, which held that Congress did not intend for section 504 to apply to treatment decisions of this sort (*United States v. University Hosp.*, 1984).

Additional litigation in the federal courts directly challenged the "final rule." In *Am. Hosp. Ass'n v. Heckler* (1984), the trial court found by summary judgment that the regulations were not authorized by the Rehabilitation Act and, thus, would be set aside. On appeal, the Supreme Court found that a hospital's withholding of treatment when no parental consent had been given did not violate the antidiscrimination provision of the Rehabilitation Act, and that the requirements for state CPS investigations posting notices, reporting, and gaining access to records were not founded on evidence of discrimination and were totally foreign to authority conferred on the secretary of DHHS (*Bowen v. Am. Hosp. Ass'n*, 1986). Furthermore the Court found that the

> Rehabilitation Act's ban on discrimination against the handicapped does not permit intervention by federal officials into medical treatment decisions traditionally left by state law to concerned parents of handicapped infants and the attending physicians, or, in exceptional cases to state agencies charged with protecting the welfare of the infant. (*Bowen v. Am. Hosp. Ass'n.*, 1986).

The Court noted that DHHS could not provide any evidence of a situation in which a hospital had denied treatment solely for reason of a handicap, nor could it provide any evidence that a hospital had failed, or had even been accused of failing, to report suspected cases of medical neglect on the part of parents to the appropriate CPS agency.

Policy Formulation, Adoption, and Implementation Stage II: Withholding Treatment as a Form of Child Abuse

While litigation was winding its way through the judicial system, Congress began to act with the passage of amendments to the Child Abuse Prevention and Treatment Act in October 1984. Disagreements over the care of newborns with disabilities held up passage of the legislation, the bulk of which dealt with more traditional issues in child abuse. In what the *Washington Post* referred to as an "act of political ecumenicism," a mixture of liberal and conservative senators as diverse as Alan Cranston and Orrin Hatch worked with the fragile coalition of 19 different organizations representing prolife and disability rights groups, and reluctant members of the medical profession[4] (Kerr, 1985).

The AAP, the American College of Obstetricians and Gynecologists, and the National Association of Children's Hospitals and Related Institutions all initially opposed the inclusion of language regarding treatment of newborns with disabilities. The AAP objected specifically to including failure to treat as a form of child abuse and instead encouraged the use of hospital review committees that had been recommended by the President's Commission. The American College of Obstetricians and Gynecologists agreed. The National Association of Children's Hospitals found it inappropriate to "address through . . . legislation the question of the appropriate medical treatment of infants . . . since other mechanisms exist to address that concern" (House of Representatives Hearings, March 9, 1983). In the end, the AAP, the American Hospital Association, and a variety of other medical groups provided support (*Congressional Q Almanac*, 1984).

The American Medical Association (AMA), however, broke with the other medical associations by mounting vocal opposition because the amendments did not adequately define the circumstances under which withholding of medically indicated treatment would be considered medical neglect. Moreover, they failed to allow for quality-of-life considerations, and were unnecessary since all states already prohibited medical neglect of children and had established procedures for dealing with such cases (AMA, 1985). Testifying against the bill, the AMA argued that:

> The natural ties of parents to their children and their love and concern for the total welfare of the infant in the context of the immediate circumstances cannot be fully understood *nor is it truly addressed by those who debate the issues in an atmosphere quite separate from the reality of the event* . . . the primary consideration should be what is best for the individual patient and not avoidance of burden to the family or to society. Quality of life is a factor to be considered in determining what is best for the individual [emphasis added]. (House of Representatives Hearings, March 9, 1983).

Edelman would have agreed that members of Congress were far removed from the complex reality and emotional trauma of the neonatal intensive care unit. They were dealing with symbols and emotions, but with little evidence of concrete facts. The AMA went on to criticize those medical associations that had participated in the negotiations and compromise. Joseph Boyle, president of the AMA, charged that "the medical groups that backed the bill had caved in for fear that if they did not, Congress or the administration would do something worse" (*Washington Post*, July 27, 1984).

Prolife and disability rights lobbyists worked strenuously to push the bill through the legislative process. Both believed that infanticide was common in the nation's NICUs. In testimony on a related bill, the Association for Retarded Citizens argued that:

> The current practice of denying medical care and nutrition to handicapped children is infanticide. Allowing such children to die because of their disability is blatantly discriminatory and possibly criminal. Physicians who suggest to parents the withholding of treatment based on future quality of life and on the child's potential lack of contribution to the family and society are greatly misleading the parents. Physicians . . . are simply not able to make such judgments when the child is one day old, one week old, or one month old (House of Representatives Hearings, Sept. 16, 1982).

Moreover, the Association joined with the National Down syndrome Congress and the American Life Lobby in opposing reliance on hospital review boards that they believed would not provide sufficient protection for newborns with disabilities. The National Right to Life Committee testified about the "growing practice of infanticide in America" and provided copies of George Will's editorial. Surgeon General Koop provided similar testimony and claimed that infanticide was not widely discussed outside a small, tight circle of those involved, but was representative of a disturbing pattern of which he was becoming increasingly aware (Senate Hearings, April 6, 11, 14, 1983).

At one point, a letter signed by both prolife and disability rights groups was mailed to every congressman. When it was rumored among prolife groups that Senator Edward Kennedy might not support the legislation, disabled citizens, some of them in wheelchairs, demonstrated in his Boston office (Paige and Karnofsky, 1986). This was a potent use of condensational symbols.

Although the coalition was fragile, it held together, and HR 1904 passed the House by an overwhelming majority (396–4) on February 1, 1984. The Senate version (S 1003) passed unanimously on July 26, 1984. The bills differed, but the conference report passed easily in September 1984.[5] Whereas DHHS had based its regulations on discrimination against the handicapped, the 1984 amendments to the Child Abuse Prevention and Treatment Act provide that failure to treat a handicapped infant falls within the meaning of child abuse and neglect (Born, 1986). Medical neglect occurs when there is a:

> failure to respond to the infant's life-threatening conditions by providing treatment [including appropriate nutrition, hydration, and medication] which, in the treating physician's reasonable medical judgment, will be most likely to be effective in ameliorating or correcting all such conditions (Child Abuse Amendments, 1984).

The amendments provide for three exceptions (that is, situations in which treatment is not medically indicated):

A. The infant is chronically and irreversibly comatose;
B. The provision of (medical) treatment would
  (i)   Merely prolong dying;
  (ii) Not be effective in ameliorating or correcting all of the infant's life-threatening conditions; or
  (iii) Otherwise be futile in terms of the survival of the infant; or

C. The provision of such treatment would be virtually futile in terms of the survival of the infant, and the treatment itself under such circumstances would be inhumane (Child Abuse Amendments, 1984).

Even if a child falls within these exceptions, appropriate nutrition, hydration, or medication must be given (Renfrow, 1984; Child Abuse and Neglect Prevention, Final Rule, 1985). Regardless of their condition or handicap, children will not be allowed to die from starvation or dehydration, nor will medication for pain be withheld. It is unclear whether other medications could be required. In making decisions whether to pursue treatment, physicians and parents can only consider medical evidence. They cannot base judgments on the kind of life the child would live or how that life might affect the family. Writing in the *Hastings Center Report,* John Moskop and Rita Saldanha argue that the law "assumes . . . that a noncomatose, nonterminal life is always preferable to nonexistence; it expressly prohibits consideration of the future quality of life of the infant" (Moskop and Saldanha, 1986).

The amendments provide that states receiving federal funds for their CPS programs are required to include medical neglect as part of their child abuse programs and procedures. CPS agencies must provide mechanisms for responding to reports of medical neglect and initiating legal proceedings when necessary to prevent the withholding of treatment. Hospitals that receive federal funds must have a designated contact person and allow CPS access to medical records. Hospitals were encouraged, but not required, to establish ICRCs.

Enforcement of the amendments lies with the states, where enforcement of other child abuse programs has been. Only states receiving federal funds for the operation of their child protective services programs are required to enforce them. If a state does not receive such funds, it does not have to implement the regulations, and the only punishment for states that do not implement them is the loss of those funds

(Murray, 1985). In 1990, California, Pennsylvania, and Hawaii did not receive funds.

The amendments grant DHHS the power to promulgate rules to implement the act. It proposed such rules in December 1984 and included a section of "clarifying" terms. These proved unacceptable to a variety of groups, including the six senators who had originally sponsored the legislation. The dispute developed over the wording of certain terms, particularly "imminent death," and also over the role ICRCs would play. The AAP threatened litigation if changes were not made. As a result, the rules were revised and issued in final form in April 1985. The word "imminent" was removed from the rules, and the role of ICRCs was more fully defined. Their role is to ensure fully informed decision making, recommend that the hospital seek CPS involvement when necessary to assure protection for the infant, educate hospital personnel and families, recommend institutional policies, and offer counsel and review for individual cases. It is not the role of the ICRC to make the decision regarding treatment for the child (Child Abuse and Neglect Prevention, Final Rule, 1985).

The 1984 Child Abuse Amendments provide a compromise solution. The language was fought out word-for-word by liberals and conservatives, prolife and disability rights groups, and physicians (Brown, 1986). The medical community would have preferred no federal intervention whatsoever, but was satisfied that physicians could continue to use "reasonable medical judgment." Prolife advocates wanted much more aggressive treatment standards. Their reasoning was that "either one believes in the sanctity of all human life, from conception till complete death (not just 'brain death'), or else one steps out, however innocently, onto a slippery slope that leads inexorably to genocide" (Murray, 1985).

The American Life Lobby felt there was too much "wiggle room" for doctors to exercise their own judgment, and Americans United for Life expressed significant reservations (Paige and Karnofsky, 1986). They were, however, satisfied that future "quality of life" could not be a consideration in deciding whether or not to treat. Disabled citizens groups, which wanted greater legal protections, were not completely happy, but thought that the debate had pointed out the value that disabled citizens add to the society and would encourage the public and the medical community to look more closely at their concerns. Finally, the Reagan administration could say that it protected the lives of helpless newborns (Rhoden and Arras, 1985).

Policy Evaluation: Was There a Problem? Is There a Change?

There has been a great deal of publicity and a significant amount of energy and money spent, but we must ask: Was there a problem to begin with, and have there been changes as a result? Policy evaluation is a difficult task, and evaluating the impact of the Child Abuse Amendments is particularly complex for a number of reasons. First, there is substantial disagreement among practitioners, ethicists, interest groups, and others who are familiar with disabled newborns, about the extent of problems prior to the promulgation of the section 504 rules and the subsequent enactment of the Child Abuse Amendments. Moreover, the language of the legislation is vague, leading to uncertainty about what the law actually requires. However, most importantly, there is inadequate information upon which to make an evaluation of policy outcomes, because there are insufficient data on the extent of medical neglect prior to the legislation, and there are relatively few studies attempting to measure the policy's impact. Additionally, even if change toward increased treatment could be identified, sophisticated analysis would be required to determine whether that change was caused by the amendments or by other factors, such as changing societal attitudes about the disabled or physicians' fears of malpractice lawsuits.

Some evidence suggests that, for the most part, care of newborns with disabilities was adequate prior to the development of the Baby Doe debate. When DHHS began to issue regulations under the 1973 Rehabilitation Act, it argued there was widespread neglect of handicapped infants (Moskop and Saldanha, 1986). However, when the Final Rule was issued, DHHS included a report on 49 cases that the Office of Civil Rights handled in 1983 before the Interim Final Rule was declared invalid. None of the reports resulted in a finding of discriminatory withholding of medical care (Nondiscrimination on the Basis of Handicap, Final Rule, 1984). The Supreme Court noted that DHHS could not provide any proof that a hospital had denied treatment solely for reasons of a handicap or had failed to report parents for medical neglect (*Bowen v. Am. Hosp. Ass'n.* 1986). In writing the rules to implement the 1984 Child Abuse Amendments, DHHS acknowledged:

> Only a very small fraction of births involve any serious question of survival. Of these, only a fraction would not be treated appropriately under current medical practice and would involve even a potential allegation of medical neglect. These considerations suggest that the potential number of cases which the statute might impact is not large (Child Abuse and Neglect, Final Rules, 1985).

The President's Commission reported in 1983 that "treatments are rarely withheld when there is a medical consensus that they would provide a net benefit to a child" (President's Commission, 1983). The AMA reached similar conclusions, writing that:

> Improper decisions regarding the withholding of medical treatment are rare, notwithstanding the complexity and uncertainty inherent in the treatment of severely ill newborns. In those rare cases where medical neglect may be suspected, state child protective

service agencies already possess the authority and capability to act (AMA, 1985).

Two recent surveys also point to the lack of need. The first is the 1987 Survey of State Baby Doe Programs conducted by DHHS. It reports that:

All States have systems in place to insure that CPS agencies can be notified immediately of any suspected instances of child abuse or neglect, including potential baby doe [*sic*] cases. *These systems have been in operation for many years* [emphasis added] (Survey of State Baby Doe Programs 1987).

As part of its study, DHHS conducted a survey of all 50 state CPS agencies. Between October 1985 and April 1987, they received a total of 19 reports of suspected medical neglect of newborns with disabilities, all of which were investigated. In only six cases did CPS intervene to change treatment. As of September 1987, one child had died despite a treatment change, a second lived in a chronic vegetative state, two were in children's convalescent hospitals, one was in foster care, and one was permanently in the custody of the state CPS agency. None of the children recovered from their disabilities, and none went home to live with their biological families.

The second survey was conducted among neonatologists and published in the *New England Journal of Medicine* (Kopelman et al., 1988). Three-quarters of the respondents believed that the Child Abuse Amendments were not necessary to protect the rights of handicapped newborns, and 81% felt they would not result in improved care for all infants. Two-thirds said the amendments interfered with parents' rights to determine the best interests of their children, and 60% indicated that the regulations did not give adequate consideration to the infant's suffering. Respondents showed some confusion about what the amendments would actually require. They also showed significant differences between what they believed the regulations required and what they would consider to be appropriate medical treatment. Overwhelming majorities (77–87%, depending on the case) wanted to consider the parents' wishes.

Although these studies point to a lack of need for the legislation, the situation is not clear-cut. For example, Fost argued in 1982 that it was common in the US to "withhold routine surgery and medical care from infants with Down syndrome for the explicit purpose of hastening death" (Fost, 1982). It is clear that, in the cases at Johns Hopkins in 1971, and Bloomington in 1982, Down syndrome children were denied surgery and therefore died. It is also clear that, in 1973, Duff and Campbell reported on the deaths of infants for whom treatment was withheld or discontinued, as did Gross and his colleagues in Oklahoma in 1983. It should also be noted that Oklahoma Children's Memorial Hospital subsequently denied that the quality-of-life formula had been a part of the selection process, or that racial and/or socioeconomic discrimination had occurred, as some had alleged. Data from the Gross research group bear out the latter. Treatment records indicate that all children on public assistance without insurance and 69% of children qualifying for medical assistance were aggressively treated, whereas 58% of patients with private insurance were aggressively treated (Civil Rights, 1989). Moreover, a higher percentage of black (100%) and Hispanic (83%) children were aggressively treated than were white (57%) children (Civil Rights, 1989).

The 1989 report of the US Commission on Civil Rights provides one of the strongest statements of need for the legislation and overall lack of enforcement. The report argues that discrimination against disabled infants occurred prior to passage of the legislation and that such discrimination continues because of lack of enforcement by government agencies, especially state CPS agencies, which make it a practice not to intervene on behalf of disabled infants (Civil Rights, 1989). The report also argues that parents are only nominally making decisions, because they are heavily influenced by physicians and other health-care practitioners who are poorly educated regarding the quality of life disabled citizens lead.

Although the Commission is to be lauded for its concern regarding the welfare of disabled citizens, it has been the center for a good deal of criticism, including condemning statements from its own chair. A number of specific criticisms are worth noting. First, the report does not adequately provide data on the incidence of medical neglect or on any changes that may have occurred since the passage of the Child Abuse Amendments. The bulk of the report is based on testimony given during two hearings. Although such testimony is interesting, it is not a substitute for quantifiable data. A number of physician opinion surveys and other studies are cited throughout the report. The overall thrust of these surveys and studies is that physicians are willing to concur with parents in withholding life-saving treatment. Although such surveys have significant value as research tools, most of them cited *predate* the passage of the Child Abuse Amendments and might show substantially different conclusions if they were conducted today.

Some of the report's findings are based on televised investigative news reports and newspaper articles. For example, the report argues: "Physicians with a bent toward denial of treatment can be quite insistent in conveying negative information, as they were with the parents of Baby Jane Doe" (Civil Rights, 1989). The report goes on to discuss physician pressure on her parents based on newspaper articles. There is no indication that an effort was made to contact the family or physicians who participated in treatment.

Much of the report's criticism of the medical profession is based on its interpretation of the role and advice given to the parents of Baby Doe by the attending obstetrician. The report argues that he was uninformed regarding the prognosis for quality of life for Down syndrome children. The Commission's statement regarding of his views may be correct or incorrect. Even if correct, one poorly informed obstetrician does not make a poorly informed profession. That is not to say that all physicians are well informed regarding physical and mental disabilities. Some research indicates the contrary (Siperstein et al., 1988). However, the Commission does not prove its case by continually referring to the events in Bloomington. Moreover, it is common practice now to transport severely impaired newborns to tertiary care centers where medical staff are more likely to be well informed.

A second criticism of the Commission is its failure to solicit the views of interested parties, especially the medical profession, and specifically physicians and hospitals who are seriously criticized in the report. Chairman Allen indicates that the Commission refused to "vent" the report among critics in advance, and therefore there was "no means to measure the sufficiency of the report's representation of the 'other side'" (Civil Rights, 1989). Indeed, a few letters from interested parties have been placed in the appendices. These parties complain of not being adequately consulted or allowed to explain their positions to the Commission.

Evaluating the impact of the Child Abuse Amendments is as problematic as trying to determine the need for them. The Kopelman study points to some change as a result of the legislation. When asked about hypothetical cases, 23–33% (depending on the case) indicated their approach to care had changed as a result of the regulations, and about half thought severely sick infants with poor prognoses were being overtreated. Thus, a substantial majority of respondents indicated their approach to treatment had not changed as a result of the legislation, and a majority indicated that overtreatment occurs. The contradiction is evident, and thus, it cannot be argued with certainty that the legislation (as opposed to other factors) caused the overtreatment. Of those indicating change, it cannot be determined accurately whether that change involved treatments *per se* or changing the process and documentation involved with decision making. The data clearly show that most respondents have not changed their approach to treatment.

Studies by Jeffrey J. Pomerance et al., on the other hand, point to significant attitudinal change between 1979 and 1985.

In 1985, despite parents' wishes, neonatologists were more willing to start and continue ventilator support in smaller and less mature infants than in 1979. In 1985, but not in 1979, respondents frequently would use ventilator support significantly more than they thought they should. Additionally, in both 1979 and 1985, it was more difficult for respondents to withhold ventilator support in infants with acquired versus congenital problems (Pomerance et al., 1988).

The authors point out that, although identical questionnaires were completed by neonatologists in both years, the composition of the study group undoubtedly changed between the two surveys. They argue that whether or not an individual respondent's views changed is moot, because the surveys show that the *collective* attitudes of those caring for newborns did change. They also point out that "it cannot be assumed that the 'Baby Doe' regulations [the Child Abuse Amendments Final Rule] were responsible for the changes in attitude, only that these changes occurred following their release. Technical, educational, or emotional factors may have influenced changes in attitudes" (Pomerance et al., 1988). The authors speculate, however, that fear of legal consequences or unwanted notoriety influenced neonatologists to use ventilator support, especially in those instances when they themselves did not think it should be used (Pomerance et al., 1988).

Another area in which change is evident, although again difficult to measure, is in the creation of ICRCs. Weir argues that, in response to the 1984 DHHS Final Rule, most hospitals having level-III NICUs and receiving federal funds established committees in response to the regulations, if they did not already have them (Weir, 1987). Studies by the AAP also show growth in the percentage of hospitals having ICRCs, although the growth is not dramatic. In 1984, 56.6% of hospitals surveyed had ICRCs, as compared to 65.7% in 1985 (Fleming et al., 1990).

Conclusion

Policy making in the United States is characterized by government fragmentation, interest-group struggle, and political symbolism. The Baby Doe debate exemplifies these characteristics. The triggering events in establishing the agenda became important symbolic tools for a nation that values human life, especially the lives of vulnerable children. Policy formulation, adoption, and implementation were dominated by the struggle of prolife and disability rights groups and medical associations. The separation-of-powers principle, with its inherent check and balance system among the president, Congress, and courts, functioned as it was designed to by the Constitution. The federal nature of the policy process was also evident as states were drawn into the enforcement phase of implementation.

However, the most important questions involve evaluation and remain unanswered: Was there a problem to begin with, and has there been change as a result of the policy enacted? Data from studies cited above and reports of specific cases related by the news media indicate that some infants died as the result of withholding treatment. It is unclear how widespread this practice was. More than likely, accurate data are unavailable. Based on the Kopelman study, it is clear that neonatologists felt the regulations were unnecessary. Information on changes in treatment practices resulting from the policy is very limited. The Kopelman data indicate that some neonatologists changed their procedures in specific cases as a result of the Baby Doe policy. The Pomerance study makes a clear statement of change, but does not conclude definitively that the Child Abuse Amendments caused the change. Both studies point to overtreatment.

Infants will always need surrogates to make decisions in their behalf. Baby Doe regulations are moving away from the trend in other areas of health care, where legal guardians and/or relatives make decisions for patients who are incapable of making decisions for themselves. How much difference is there between a parent and physician who jointly make a decision for a critically ill newborn, and an adult child and physician who make a similar decision for a critically ill, mentally incompetent parent? The latter is standard procedure and typically not questioned.

Much of what went on in the political arena was symbolic. Again, Edelman is insightful:

> There is no check on the fantasies and conceptualizing of those who can never test objectively their conviction(s) . . . nor is there the check of reality and feedback. Conclusive demonstrations that their heroes' policies may often be futile or misconceived are impossible simply because the link between dramatic political announcements and their impact on people is so long and so tangled (Edelman, 1964).

Treatment of newborns with disabilities is an extraordinarily complex task. There are no simple answers, but as Edelman has pointed out, condensational symbols allow people to simplify and distort. Reason and analysis are gone, and emotions are fueling the debate. The Baby Doe debate exemplifies this symbolism in that it reflects the participants' feelings rather than the complexities with which medical staff and families must deal. In this regard, the political arena is a far distance from the nation's NICUs.

Notes

[1] Generally speaking, this chapter refers to infants born with physical or mental impairments as "newborns with disabilities." The term "handicapped" is used in reference to regulations or court decisions that adopted that language.

[2] Although it has been alleged that the parents and physicians were attempting to starve the infants to death, this may not have been the case. It is reported that the medical staff initially believed that the infants had a trachea esophageal fistula, which would have caused food taken orally to run into the lungs and thereby hasten death (Staver, 1981).

[3] The signatories to the joint statement included: The American Academy of Pediatrics, the Association for Retarded Citizens, the National Association of Children's Hospitals and Related Institutions, Inc., the Spina Bifida Association of America, the Association for Persons with Severe Handicaps, the American Association on Mental Deficiency, the American Association of University Affiliated Programs for Persons with Developmental Disabilities, the Coalition of Citizens with Disabilities, Inc., and the National Down Syndrome Congress.

[4] Senate sponsors included Orrin G. Hatch (R-UT), Alan Cranston (D-CA), Jeremiah Denton (R-AL), Christopher J. Dodd (D-CT), Nancy Landon Kassebaum (R-KS), and Don Nickles (R-OK). Interest groups included: The National Right to Life Committee, the American Life Lobby, the National Association of Children's Hospitals, the American Academy of Pediatrics, the American Nurses Association, and a variety of advocacy groups for the handicapped. *See* "Senate passes bill to require care of deformed babies," *Washington Post*, July 27, 1984.

[5] The House version would have required more heroic measures on the part of physicians. The conference report was based primarily on the Senate version. *See* Child abuse 'Baby Doe' legislation cleared. (1984) *Congressional Quarterly Almanac*.

Acknowledgment

Portions of this chapter are reprinted with the permission of the Policy Studies Organization and Greenwood Press.

References

American Academy of Pediatrics, Committee on Bioethics (1983) Treatment of critically ill newborns. *Pediatrics* **72**, 565,566.

American Academy of Pediatrics v. Heckler, 561 F. Supp. 395 (D.D.C. 1983).

American Hospital Association v. Heckler, 585 F. Supp. 541 (S.D.N.Y. 1984).

American Medical Association (1985) Statement Made to the US Commission on Civil Rights.

Anderson, J. E. (1990) *Public Policy-Making,* "The Study of Public Policy" (35); "Policy Formation" (83); "Policy Implementation" (191–195); "Policy Impact, Evaluation, and Change" (223,224) (Houghton Mifflin, Boston).

Annas, G. J. (1986) Checkmating the Baby Doe regulations. *Hastings Cent. Rep.* **16,** 29–31.

Born, M. A. (1986–87) Baby Doe's new guardians: Federal policy brings nontreatment decisions out of hiding. *Kentucky Law J.* **75,** 659–675.

Bowen v. American Hospital Association, 106 S. Ct. 2101 (1986).

Brown, L. D. (1986) Civil rights and regulatory wrongs: The Reagan administration and the medical treatment of handicapped infants. *J. Health Politics, Policy and Law* **11,** 231–254.

Child Abuse Amendments of 1984, Pub. L. No. 98-457, 121, 98 Stat. 1749 (1984).

Child abuse, "Baby Doe" legislation cleared. (1984) *Congressional Quarterly Almanac* 482–484.

Cobb, R. W. and Elder, C. D. (1972) *Participation in American Politics: The Dynamics of Agenda-Building,* "Issue Creation and Agenda Content" (85) (Allyn and Bacon, Boston).

Department of Health and Human Services (1982) Notice to Health Care Providers. 47 Fed. Reg. 26,027 (June 16, 1982).

Department of Health and Human Services (1983) Nondiscrimination on the Basis of Handicap, Interim Final Rule, 48 Fed. Reg. 9630 (modifying 45 C.F.R. 84.61).

Department of Health and Human Services (1984) Nondiscrimination on the Basis of Handicap, Procedures and Guidelines Relating to Health Care of Handicapped Infants, Final Rule, 49 Fed. Reg. 1622 (codified at 45 C.F.S. 84.55).

Department of Health and Human Services (1985) Child Abuse and Neglect Prevention and Treatment Program, Final Rule, 50 Fed. Reg. 14878 (codified at 45 C.F.R. Part 1340).

Department of Health and Human Services (1987) Survey of State Baby Doe Programs.

Doe v. Bloomington Hospital, 104 S.Ct. 394 (1983).

Duff, R. and Campbell, A. G. M. (1973) Moral and ethical dilemmas in the special-care nursery. *N. Engl. J. Med.* **289,** 890–894.

Edelman, M. (1964) *The Symbolic Uses of Politics,* "Introduction" (7); "Symbols and Political Quiescence" (31); "Language and the Perception of Politics" (124–125) (University of Illinois Press, Chicago).

Fleming, G. V., Hudd, S. S., LeBailly, S. A., Greenstein, R. M. (1990) Infant care review committees. *Am. J. Dis. Child.* **144,** 778–781.

Fost, N. (1982) Passive euthanasia of patients with Down's syndrome. *Arch. Int. Med.* **142,** 2295.

Fost, N. (1986) Treatment of seriously ill and handicapped newborns. *Crit. Care Clin.* **42:2,** 149–159.

Gross, R. H., Cox, A., Tatrek, R., Pollay, M., Barnes, W. A. (1983) Early management and decision-making for the treatment of myelomeningocele. *Pediatrics* **72,** 450–458.

Joint Policy Statement (1984) Principles of treatment of disabled infants. *Pediatrics* **73,** 559,560.

Kerr, K. (1985) Negotiating the compromise. *Hastings Cent. Rep.* **15,** 6,7.

Kopelman, L. M., Irons, T. G., and Kopelman, A. (1988) Neonatologists judge the "Baby Doe" regulations. *N. Engl. J. Med.* **318,** 677–683.

Moskop, J. C. and Saldanha, R. L. (1986) The Baby Doe rule: Still a threat. *Hastings Cent. Rep.* **16,** 8–14.

Murray, T. H. (1985) The final, anticlimactic rule of Baby Doe. *Hastings Cent. Rep.* **15,** 5–9.

Nelson, B. J. (1984) *Making an Issue of Child Abuse: Political Agenda Setting for Social Problems,* "Theoretical Approaches to Agenda Setting" (20) (University of Chicago Press, Chicago).

Paige, C. and Karnofsky, E. B. (1986) The antiabortion movement and Baby Jane Doe. *J. Health Politics, Policy and Law* **11,** 255–269.

Pomerance, J., Yu, T. C., and Brown., S. J. (1988) Changing attitudes of neonatologists toward ventilator support. *J. Perinatol.* **8:3,** 232–241.

President's Commission for the Study on the Ethical and Legal Problems in Medicine and Biomedical and Behavioral Research (1983) Deciding to Forego Life-Sustaining Treatment: A Report on the Ethical, Medical, and Legal Issues in Treatment Decisions, pp. 217, 225, 226.

Rehabilitation Act of 1973, 504, 29 U.S.C. 794 (1982).

Renfrow, J. F. (1984) The Child Abuse Amendments of 1984: Congress is calling North Carolina to respond to the Baby Doe dilemma. *Wake Forest Law Rev.* **20,** 975–1000.

Rhoden, N. K. and Arras, J. D. (1985) Withholding treatment from Baby Doe: From discrimination to child abuse. *Milbank Memorial Fund Q.* **63,** 18–51.

Shaw, A. (1977) Defining the quality of life. *Hastings Cent. Rep.* **7:5,** 11.

Shaw, A., Randolph, J. G., and Manard, B. (1977) Ethical issues in pediatric surgery: A national survey of pediatricians and pediatric surgeons. *Pediatrics* **60,** 588–599.

Siperstein, G., Wolraich, M. L., Reed, D., and O'Keefe, P. (1988) "Medical decisions and prognostications of pediatricians for infants with meningomyelocele." *J. Pediatr.* **113,**835–840.

Staver, S. (1981) Siamese twins' case "devastates" MDs. *Am. Med. News* **9:1,** 47,48.

Todres, D., Krane, D., Howell, M., and Shannon, D. (1977) "Pediatricians' attitudes affecting decision-making in defective newborns." *Pediatrics* **60:2,** 197–201.

(1976) Treating the defective newborn. *Hastings Cent. Rep.* **6:2,** 2.

United States v. University Hospital of the State University of New York at Stony Brook, 575 F. Supp. 607 (1983).

United States v. University Hospital, State University of New York at Stony Brook, 729 F. 2d. 144 (1984).

US Commission on Civil Rights (1989) *Medical Discrimination Against Children with Disabilities,* pp. 3,28,70,71,81,136–138,149,155,356,367.

US House of Representatives (1983) Subcommittee on Select Education. Hearings on HR 1904.

US House of Representatives (1982) Subcommittee on Select Education. Hearings on Treatment of Infants Born with Handicapping Conditions, p. 52.

US Senate (1983) Subcommittee on Family and Human Services. Hearings on Child Abuse Prevention and Treatment and Adoption Reform Act Amendments of 1983; 7, 49.

Washington Post. Baby Doe task force hits snag. May 23, 1983.

Washington Post. Handicapped groups back "Baby Doe" rule. March 30, 1983.

Washington Post. Handicapped policy undergoing a rewrite. March 4, 1982.

Washington Post. DHHS to publish compromise "Baby Doe" rule. January 6, 1984.

Washington Post. Koop acted as midwife for new "Baby Doe" rule. January 10, 1984.

Washington Post. Senate passes bill to require care of deformed babies. July 27, 1984.

Weber v. Stony Brook Hospital, 467 N.Y.S. 2d 687 (A.D. 2 Dept. 1983).

Weber v. Stony Brook Hospital, 469 N.Y.S. 2d 65 (Ct. App. 1983).

Weir, R. (1984) *Selective Nontreatment of Handicapped Newborns* (Oxford University Press, NY), p. 960.

Weir, R. (1987) Pediatric Ethics Committees: Ethical advisers or legal watchdogs? *Law, Medicine, and Health Care* **15:3,** 99–109.

Will, G. The killing will not stop. *Washington Post.* April 22, 1982.

A Legal Analysis
of the Child Abuse
Amendments of 1984

James Bopp, Jr. and Mary Nimz

In April 1982, a child called Baby Doe was born in
Bloomington, Indiana with a surgically correctable, but life-
threatening, condition that prevented oral feeding (Bopp,
1985). The child also had Downs syndrome (Bopp, 1985). Over
objections from the child's pediatricians, the mother's obstetrician counseled the parents to refuse all care and treatment
for the infant, including surgery to correct the feeding problem and all forms of food and water. The obstetrician testified that he had told the parents that they had the
"alternative" of "do[ing] nothing" and the child would probably live only a matter of several days. The obstetrician based
his recommendation on the belief that, even with successful
surgery, the child would still have Downs syndrome and
therefore "the possibility of a minimally adequate quality of
life would be non-existent."

The parents followed the obstetrician's advice and
refused consent to food, water, and medical care. Nurses
at the hospital protested Baby Doe's lack of care, which

From: *Compelled Compassion* Eds.: Caplan, Blank, and Merrick
©1992 The Humana Press Inc.

prompted the hospital to seek a court order. The court directed Bloomington Hospital to follow the obstetrician's order, and the Indiana Supreme Court refused to review the case. In the meantime, families came forward with offers to adopt and care for Baby Doe. Lawyers representing the child's interest were en route to Washington, DC to file a petition for review with the United States Supreme Court when Baby Doe died six days after birth without having received any food, water, or medical care (Bopp, 1985).

Baby Doe's death became a catalyst for this nation's ethical, medical, and legal debate on how our society should treat infants born with disabilities. This debate has attempted to solve the conflicts between the infant's rights, the parents' decision-making rights, and the state's interests and *parens patriae* authority in this area. Legislative and administrative responses have attempted to provide protections for infants with disabilities. The most comprehensive solution offered thus far has been the Child Abuse Amendments of 1984 (Child Abuse Amendments, 1984). In 1989, the Child Abuse Amendments were reauthorized (Child Abuse Reauthorization, 1989).

Brief Related History

At the administrative level, in direct response to the death of Baby Doe, President Reagan directed the Department of Health and Human Services (DHHS) to notify healthcare providers of the applicability of Section 504 of the Rehabilitation Act of 1973. The regulations issued by DHHS required posting a notice in hospital wards—likely to treat newborns with disabilities—stating that Section 504 prohibited hospitals receiving federal financial assistance from withholding nutrition, medical, or surgical treatment from handicapped newborns (DHHS, 1983a). DHHS' Interim Final Rule published to expedite enforcement of Section 504 was enjoined by a federal district court as "arbitrary and capricious" (*American Academy of Pediatrics v. Heckler*, 1983). The court found insufficient evidence of an emergency that would warrant waiving the 60-day comment period required by the Administrative Procedures Act (APA).

Subsequently, the parents of Baby Jane Doe, an infant born with multiple disabilities on Long Island, New York, refused consent to surgical procedures for her. After failed attempts by DHHS to obtain Baby Jane Doe's medical records, the United States brought an action in federal district court to compel the hospital to make the infant's medical records available to the government on the basis that the hospital, a recipient of federal financial assistance, had violated the procedural regulations promulgated under Section 504 (*United States v. University Hosp., SUNY at Stony Brook,* 1983). The court held that the hospital had not violated Section 504 because it was the parents, not the hospital, that had refused consent to surgery, and that the parents' refusal was reasonable because it was based on due consideration of available medical options (*United States v. University Hosp., SUNY at Stony Brook,* 1983). The United States Court of Appeals for the Second Circuit affirmed the lower court's decision, but stated that the DHHS investigation and request for access to medical records were not within the scope of Section 504 (*United States v. University Hospital, SUNY at Stony Brook*, 1984).

On January 12, 1984, DHHS issued a Final Rule under Section 504 (DHHS, 1984a). These regulations reduced the size of the notice to be posted in hospitals and required its posting only in nurses' stations. More significantly, it required state child protective service agencies receiving federal financial assistance to develop procedures for dealing with reports of denial of treatment to infants because they are disabled. The regulations allowed for on-site investigations, and authorization to seek court orders to compel the provision of necessary nutrition and medical treatment. Although the 16,331 comments received were nearly unanimously supportive of the proposed regulations, medical groups voiced strong opposition.

The Final Rule was struck down on June 11, 1984, as invalid and unlawful under the APA, because it was promulgated without statutory authority (*Am. Hosp. Ass'n. v. Heckler*, 1984), and the Second Circuit Court of Appeals affirmed. Despite legislative history indicating otherwise (Senate Report, 1974), both courts felt bound by the ruling in *University Hospital* that held that "Congress never contemplated that Section 504 would apply to treatment decisions of this nature" (*United States v. University Hospital, SUNY at Stony Brook,* 1984).

On review, the United States Supreme Court affirmed the Second Circuit's decision in a plurality opinion. The plurality held that, in situations in which parents withhold consent to treatment for their disabled child, the child is not "otherwise qualified" to receive treatment and, therefore, Section 504 is not violated (*Bowen v. Am. Hosp. Ass'n.,* 1986).

The Negotiation Process of the Child Abuse Amendments of 1984

Although DHHS attempted to provide protection for infants with disabilities, the Congress was also considering its role in this issue. Legislators felt that it had become necessary, in light of the trend of denying treatment to infants with disabilities that surfaced in the cases of Baby Doe and Baby Jane Doe, to set into law the minimum standard of care required for the treatment of infants with disabilities. Legislative action began shortly after Baby Doe's death. Representative John Erlenborn introduced a bill that became the catalyst for hearings and committee action that ultimately led to the enactment of the Child Abuse Amendments of 1984. The final version of the Child Abuse Amendments of 1984 resulted from a lengthy negotiation process that included give and take among disability groups, medical associations, right-to-life organizations, and a bipartisan group of senators whose usual positions varied dramatically.

In midApril 1984, Senator Denton of Alabama set up an advisory committee to study the issue of medical neglect of infants born with disabilities (Gerry and Nimz, 1987). Physicians, but no advocacy groups, were called to discuss the issue with senatorial staff aides and personnel from the Senate Committee on Labor and Human Resources.

Subsequently, Senator Dodd's staff began meeting with medical groups to explore the issues surrounding treatment for infants with disabilities. At these meetings, the opinions offered by physicians revealed a bias toward compromising infants' lives through the use of "quality-of-life" decision making. Moreover, it became apparent that hospital review committees set up at that time to evaluate treatment decisions for infants with disabilities held no room for committee members who dissented from nontreatment decisions. At this juncture, Democratic senatorial staff aides came to the conclusion that specific guidelines were needed for dealing with the medical treatment of infants with disabilities.

In an attempt to gain diversified input, Senator Nickel's staff aide Laura Clay arranged a showing of the Cable News Network documentary, "Baby Doe Special Assignment." In April 1984, Senator Kennedy's staff prepared a draft of proposed language dealing with protections for infants with disabilities from medical neglect (Senate Bill 1003, April 1984a). This draft proposed that, in exchange for federal child abuse funds, each state create an ombudsman program for the purpose of protecting and advocating for the interests of infants born with life-threatening impairments and infants or children with "special needs."

The medical community rejected the idea of an ombudsman program as an intrusion into their area of expertise. In response, the American Academy of Pediatrics proposed a plan to be used by state child protective service agencies to respond to reports of unlawful medical neglect, including instances of inappropriate withholding of medical treatment from infants at risk with life-threatening congenital impairments (Senate Bill 1003, April 1984b).

A second Kennedy staff draft of proposed amendments was released in late April 1984 (Senate Bill 1003, April 1984c). This draft proposed that the secretary of the Department of Health and Human Services publish "regulations for the establishment of local decision-making procedures within each health care facility as to medically indicated treatment of newborns with congenital impairments who are at risk with a life-threatening condition," and for the implementation of subparagraph (K) "which shall at a minimum require that all such newborns be provided treatment provided to other similarly situated children without congenital impairments that is necessary to correct or ameliorate a life-threatening condition, relief from suffering including feeding, and medication for pain and sedation as appropriate" (Senate Bill 1003, April 1984c). Also in late April, Senator Cranston's staff began meeting with disability rights and prolife groups separate from its meetings with medical groups.

On May 18, 1984, a Senate Committee on Labor and Human Resources staff discussion draft of proposed legislation was released (Senate Bill 1003, May 18, 1984). This draft still contained the ombudsman program, but provided a new definition of "medically indicated treatment" as "treatment which would normally be provided to infants without regard to the presence of disabling conditions and includes treatment specifically designed to ameliorate a disabling condition" (Senate Bill 1003, May 18, 1984). Again, separate meetings were held by the Kennedy staff with the medical groups, and with the disability rights and prolife groups. On May 22, 1984, the disability rights groups and prolife organizations offered a critique of the ombudsman program, outlining its weaknesses. Areas of concern included the selection process for choosing an ombudsman, the ombudsman's access to medical records, the standard of care language, and the time-frame in which the legislation would go into effect (Senate Bill 1003, May 22, 1984).

On June 7, 1984, a second staff discussion draft, which reflected the comments received from the staff's meetings with the medical, disability rights, and prolife groups, was prepared (Senate Bill 1003, June 7, 1984). In this draft, the ombudsman program was dropped, and instead, Section (K) required states that receive federal grant money for child abuse and neglect agencies to have in place "state procedures or programs to respond to instances of inappropriate withholding of medically indicated treatment from disabled infants or children with life-threatening conditions" (Senate Bill 1003, June 7, 1984). In addition, the definition of medically indicated treatment was redefined to mean "the type of treatment which would be provided to other similarly situated infants or children with or without a disabling condition and includes the provision of appropriate nutrition, hydration, and medication" (Senate Bill 1003, June 7, 1984). The staff discussion draft defined the term "local decision-making entity" as "a committee or other group of individuals, within a health-care facility, that is organized for the purpose of reviewing the treatment or the withholding or termination of treatment for a disabled infant or child with a life-threatening condition" (Senate Bill 1003, June 7, 1984).

In a June 13 response draft from prolife organizations, an alternative definition for "local decision-making entity" was offered (Senate Bill 1003, June 13, 1984). It provided that "the term 'local decision-making entity' means a committee or other group of individuals, within a health-care facility, which has the responsibility to insure that disabled infants and children with life-threatening conditions receive medically indicated treatment" (Senate Bill 1003, June 13, 1984).

After two months of accepting comments on the staff drafts, the staff aides brought all the groups together for face-to-face negotiations for the first time on or shortly after June 19, 1984. These negotiations lasted for 50 hours during a one-week period. The medical groups at the negotiations included the American Academy of Pediatrics, the American Hospital Association, the American Academy of Obstetricians and Gynecologists, the California Association of Children's Hospitals, the American College of Physicians, and the American Association of Medical Colleges. The disability rights groups represented included the Association for Retarded Citizens of the United States, the Spina Bifida Association of America, the Down Syndrome Congress, and The Association for Persons with Severe Handicaps. The prolife organizations participating were the National Right to Life

Committee and the Christian Action Council. Susanne
Martinez, from Senator Cranston's staff, organized the
meetings, and Jon Steinberg, also from Senator Cranston's
staff, was selected to act as arbitrator between Martin
Gerry—who chaired the coalition between disability rights
groups and prolife organizations—and the American Academy of Pediatrics, the main negotiator from the medical
groups. During these face-to-face sessions, various terms and
definitions were offered by both sides until consensus language was agreed upon.

During the face-to-face negotiations, a proposed draft of
language supported by the medical groups was used as the
starting point (Senate Bill 1003, June 19, 1984). The medical groups offered this initial definition: "inappropriate withholding of medical treatment means the failure to provide
medical or surgical treatment, including treatment specifically designed to ameliorate a disabling condition, when such
failure is clearly contrary to the best interests of the infant"
(Senate Bill 1003, June 19, 1984). The disability rights groups
and prolife organizations found the definition of the term
"inappropriate" as "contradictory," because "it might indicate
that different considerations are in order when dealing
with a handicapped child" (Senate Bill 1003, June 14, 1984).
Moreover, "since the bill already defines 'medically indicated
treatment,' it was believed that insertion of the word 'inappropriate' would only create ambiguity where none was
intended" (Senate Bill 1003, June 13, 1984). The disability
rights groups and prolife organizations felt that the withholding of any treatment that is medically indicated would be
inappropriate. Consequently, the term inappropriate was
deleted from the definition. Furthermore, the disability rights
groups and prolife organizations found the medical groups'
definition to be unacceptable in that it contained "best-interests" language, which suggested treatment may be withheld
based on a quality-of-life assessment. This terminology was
also deleted upon agreement by all groups.

Language similar to that of an earlier Kennedy staff
draft was also proposed. In this draft, the "term 'withhold-

ing of medically indicated treatment' means the failure to provide the treatment which would be provided to other similarly situated infants and children with or without a disabling condition in order to correct or ameliorate a life-threatening condition, and includes the failure to provide appropriate nutrition, hydration, and medication." This language was subsequently rejected because of possible equal-protection problems. Under this language, a disabled infant could still be denied treatment based on disability, because he or she is not similarly situated to a nondisabled infant needing the same treatment.

Standard of Care

After several transformations, a basic definition was agreed upon and became the language adopted by Congress. The term " 'withholding of medically indicated treatment' means the failure to respond to the infant's life-threatening conditions by providing treatment [including appropriate nutrition, hydration, and medication] which in the treating physician's or physicians' reasonable medical judgment, will be most likely to be effective in ameliorating or correcting all such conditions" (Child Abuse Amendments, 1984).

DHHS summarized the mandated standard of care as follows:

> [F]irst, all such disabled infants must under all circumstances receive appropriate nutrition, hydration and medication. Second, all such disabled infants must be given medically indicated treatment. Third, there are three exceptions to the requirement that all disabled infants must receive treatment, or, stated in other terms, three circumstances in which treatment is not considered "medically indicated" (DHHS, 1984b).

The regulations make clear that "appropriate nutrition, hydration and medication" must be provided to such infants under all circumstances (DHHS, 1983b). This provision provides clear guidance on an increasingly controversial issue.

A number of ethicists, physicians, and courts have opined that there is no distinction between provision of medical treatment and provision of nutrition and hydration. They argue that food and water may sometimes be withheld in situations in which surgery or other forms of medical treatment are withheld (Paris and Fletcher, 1983; Lynn and Childress, 1983). For court decisions, *see In re Conroy*, 1985 and *Barber v. People*, 1983. Others argue that intravenous and tube feeding are generally less intrusive and burdensome as compared to surgery or chemotherapy. Feeding by artificial means is more closely aligned with nursing care and other forms of nonheroic, supportive care that should be provided to all patients even if they are dying (Derr, 1986).

Some commentators have asserted that "nutrition, hydration, and medication" are not "appropriate" in all cases (Paris and Fletcher, 1983). However, legislative history plainly rejects such an interpretation of the language of the Act. One of the questions asked on the Senate floor, before the adoption of the compromise language, was whether "the words 'appropriate nutrition [and] hydration,' as used throughout the amendment, are not meant to sanction outright denial of all nutrition and hydration, but are intended only to affirmatively require appropriate nutrition and hydration in all cases. In other words, nothing in this amendment allows an infant to be denied nutrition and hydration." The answer by the floor leader for the amendment was, "Yes, that is correct" (Congressional Record, 1984a) (colloquoy between Senators Helms and Hatch). A Joint Explanatory Statement provided: "[t]he six principal sponsors of this compromise measure [intend] that this statement . . . be the definitive legislative history in the Senate on it. Any remarks of individual views including the principal sponsors, on this legislation express only their personal views and do not, therefore, constitute authoritative interpretation or explanation of the measure" (Congressional Record, 1984b).

The Interpretative Guidelines issued with the Final Rules provided that "it should be clearly recognized that the statute is completely unequivocal in requiring that all infants receive 'appropriate nutrition, hydration, and medication,' regardless of their condition or prognosis" (DHHS, 1985a). Thus, although some physicians might consider total withdrawal of nutrition and hydration "appropriate," even when it could be effectively assimilated by a dying or disabled patient, this view is rejected by the Act and regulations.

The Act and regulations, therefore, make it clear that "comfort care" or "palliative treatment" is the baseline that must be provided to every patient, even when efforts at curing or reversing a disease process have been abandoned. This comfort care must include basic sustenance in the form "appropriate" to the patient's condition, e.g., patients incapable of receiving food and liquids orally might most appropriately receive nutrition and hydration through such measures as intravenous fluids, nasogastric or gastric tube feedings, or hyperalimentation. It also includes medication needed to make the patient comfortable, such as painkillers and drugs necessary to cope with transient conditions that decrease the comfort of the patient, e.g., a urinary tract infection.

"Medication" should be construed in congruence with the "comfort care" concept. Obviously, Congress did not mean to imply that all conceivable medication must always be given; it intended that medication "appropriate" to the comfort care baseline be required. Similarly, it is obvious that Congress, having very specifically spelled out precise conditions under which treatment to correct or ameliorate life-threatening conditions should be provided or could be withheld, did not intend, through use of the word "appropriate," to give a standardless discretion to physicians to withhold basic comfort care altogether.

The Interpretative Guidelines go so far as to maintain that the palliative care baseline invariantly required by the Act encompasses some treatment beyond medication. Regarding "palliative treatment to make a condition more tolerable" as treatment "that will, in the treating physician's reasonable medical judgment, relieve severe pain . . . ," DHHS took

note that "medication [is] one [but not the only] potential palliative treatment to relieve severe pain . . ." (DHHS, 1985a).

It is DHHS' interpretation that the term "not be effective in ameliorating or correcting all of the infant's life-threatening conditions" does not permit the withholding of
ameliorative treatment that, in the treating physician's or
physicians' reasonable medical judgment, will make a condition more tolerable (such as providing palliative treatment
to relieve severe pain, even if the overall prognosis, taking
all conditions into account, is that the infant will not survive) (DHHS, 1985a).

Following the language of the statute, the regulations
define "withholding of medically indicated treatment" as "the
failure to respond to the infant's life-threatening conditions
by providing treatment which [including appropriate nutrition, hydration, and medication], in the treating physician's
[or physicians'] reasonable medical judgment, will be most
likely to be effective in ameliorating or correcting all such
conditions . . . " (DHHS, 1983b). "[R]easonable medical judgment" is defined as "a medical judgment that would be made
by a reasonably prudent physician, knowledgeable about the
case and the treatment possibilities with respect to the medical conditions involved" (DHHS, 1983b).

In the Interpretative Guidelines, DHHS interprets the
term "treatment" to include diagnostic procedures required
to evaluate the need for particular medical intervention,
including, when necessary, "further evaluation by, or consultation with, a physician or physicians whose expertise is
appropriate to the condition[s] involved or further evaluation at a facility with specialized capabilities regarding the
condition[s] involved." The term "treatment" is also interpreted to mean not merely "a particular medical treatment
or surgical procedure," but rather "a complete potential treatment plan," including "multiple medical treatments and/or
surgical procedures over a period of time designed to ameliorate or correct a life-threatening condition or conditions"
(DHHS, 1985a).

The Exceptions

After the definition of "withholding of medically indicated treatment" was agreed upon by the negotiating parties, attention was focused on under which circumstances the failure to provide treatment (other than nutrition, hydration, and medication) would be considered medical neglect. This discussion did not involve nutrition, hydration, and medication, since it was agreed that, under all circumstances—including the exceptions—nutrition, hydration, and medication must be provided to the infants. To clarify these requirements, treatment was defined as relating to life-threatening conditions, whereas medication relates to other conditions as well and includes, but is not limited to, antibiotics and sedation. The proposed language "pain, suffering or other distress" was rejected in favor of the phrase "other than nutrition, hydration and medication for pain, suffering and other distress" to give effect to such meaning.

Some medical groups proposed the category of premature infants or low-birthweight infants as an exception from the definition of withholding medically indicated treatment (Gerry and Nimz, 1987). This proposal was found to be unacceptable in that it was too broad. Many infants in this category have a chance for survival and would benefit from treatment. Another proposed exception by medical groups would have included infants who had a diagnosis or prognosis for being profoundly and severely mentally retarded. The use of this criterion was found to be an illegitimate reason for withholding medically indicated treatment and was rejected (Nerney, 1987).

The first exception of the Child Abuse Amendments provides that the term "withholding of medically indicated treatment" does not include failure to provide treatment (other than nutrition, hydration, and medication) to an infant when, in the treating physician's or physicians' reasonable medical judgment, "the infant is chronically and irreversibly comatose" (Child Abuse Amendments, 1984).

The predecessor language proposed by the medical groups for this exception was "if the infant is permanently and completely unconscious" (Senate Bill 1003, June 25, 1984). This phrase was explicitly rejected as being too broad and was subsequently replaced with the term "comatose." After additional concerns were raised that the word "comatose," standing alone, was also too broad, the adjectives "chronically and irreversibly" were added. All the medical groups, except the American Medical Association, agreed to the use of this phrase.

The second exception allows nontreatment (other than nutrition, hydration, and medication) when the provision of such treatment would "(a) merely prolong dying, (b) not be effective in ameliorating or correcting all of the infants' life-threatening conditions, or (c) otherwise be futile in terms of the survival of the infant" (Child Abuse Amendments, 1984). Proposed language for this exception included the phrase "prolong the act of dying." During discussion, this term was shown to be too ambiguous, because it did not convey a sufficiently definite sense of time.

The Interpretative Guidelines contain extensive commentary on the meaning of terms in this definition (DHHS, 1985a). First, DHHS interpreted the words "merely prolong dying." It affirmed that the language did not "apply where many years of life will result from the provision of treatment, or whether the prognosis is not for death in the near future, but rather the more distant future" (DHHS, 1985a). Thus, treatment may not be withheld from any and every infant diagnosed as terminally ill. The child's death must be expected to occur relatively soon, even if treatment is provided, in order for the exception to apply. At the same time, DHHS, at the urging of senators who had drafted the compromise statutory provisions, withdrew its originally proposed requirement that, in order for the exception to apply, it must be the case that "death is imminent and treatment will do more than postpone the act of dying" (DHHS, 1984b). Instead, it wrote, "[t]he Department makes no effort to draw an exact line to separate 'near future' from 'more distant future.'" That determination is to be made in accordance with "reasonable medical judgment" (DHHS, 1985a).

The second subcategory of the second exception was agreed upon because it defines a circumstance in which treatment is futile. Since there are other circumstances where treatment is futile, the third subcategory was added. The negotiations were careful, however, to add the phrase "in terms of the survival of the infant" to emphasize that it is the death of the infant that cannot be prevented, not that the treatment will fail to restore the infant to full health and functioning.

The last exception excludes from the term "withholding of medically indicated treatment" when "the provision of such treatment would be virtually futile and the treatment itself under such circumstances would be inhumane" (Child Abuse Amendments, 1984). This exception underwent considerable transformation before consensus language was agreed upon. The original proposal by the medical groups was to exclude treatment "where the provision of treatment would be inhumane or it would be unconscionable to provide treatment" (Gerry and Nimz, 1987). The inclusion of the term "unconscionable" was explicitly rejected, because it shifts the focus from the effectiveness of the treatment for the infant to the motive of the person who provides treatment. Moreover, the word "unconscionable" suggests that to withhold treatment based on the infant's quality of life would be acceptable. In addition, the term "inhumane" was unacceptable, because it would allow no treatment of some infants who had a chance for survival, but whose quality of life would be low. The "inhumane" language was accepted only after it was coupled with the language that the treatment would be "virtually futile," thus, only allowing withdrawal of treatment on "humane" grounds when there was almost no chance of survival for the infant. Throughout the negotiations, the American Medical Association repeatedly sought an exception relating to pain, such as "excessive pain to the infant" or "if pain cannot be abolished or alleviated" (Gerry and Nimz, 1987). These proposals were rejected with the argument that treatment should not be withheld from a nondisabled infant on the basis of pain.

Other classes of the statutory definition, DHHS wrote in its Interpretative Guidelines, focus on the expected result of the possible treatment. This provision of the statutory definition adds a consideration relating to the process of possible treatment. It recognizes that, in the exercise of reasonable medical judgment, there are situations, where although there is some slight chance that the treatment will be beneficial to the patient (the potential treatment is considered virtually futile, rather than futile), the potential benefit is so outweighed by negative factors relating to the process of treatment itself that, under the circumstances, it would be inhumane to subject the patient to the treatment (DHHS, 1985a).

The importance of the term "inhumane" is affected by a sentence in the Joint Explanatory Statement by Principal Sponsors of Compromise Amendment Regarding Services and Treatment for Disabled Infants, which was intended to serve in place of a committee report:

> The use of the term "inhumane" in exception (C), above, is not intended to suggest that consideration of the humaneness of a particular treatment is not legitimate in any other context; rather, it is recognized that it is appropriate for a physician, in the exercise of reasonable medical judgment, to consider the factor in selecting among effective treatments (Congressional Record, 1984b).

This language of the Act and the Joint Explanatory Statement might be subject to abusive distortion were it not for the very clear statements made by DHHS in the course of promulgating the Final Regulations.

At the conclusion of the week-long negotiations, a consensus was obtained on the language among the medical groups, the disability rights groups, and the prolife organizations. The consensus bill was supported by the American Hospital Association, Catholic Health Association, National Association of Children's Hospitals and Related Institutions,

American Academy of Pediatrics, American College of Obstetricians and Gynecologists, American Nurses Association, American College of Physicians, California Association of Children's Hospitals, Nurses Association of American College of Obstetricians and Gynecologists, American Association of Mental Deficiency, Association for Retarded Citizens, US, Spina Bifida Association of America, National Down Syndrome Congress, People First of Nebraska, Association for Persons with Severe Handicaps, Disability Rights Center, Operation Real Rights, Christian Action Council, National Right to Life Committee, and the American Life Lobby (House Conference Report, 1984). The American Medical Association did not endorse it.

In fact, the day after the consensus was reached, the American Medical Association again insisted that the bill must allow nontreatment of infants in cases in which treatment would be inhumane and unconscionable. During one negotiating session, a spokeswoman for the AMA said this might apply in a case in which medical treatment would save the infant's life, but the infant would subsequently live in pain (Gerry and Nimz, 1987).

The AMA's proposal was completely unacceptable to the coalition of prolife and disability rights groups. Moreover, a staff member to one of the Democratic senators involved in the negotiations commented that the AMA language would amount to "programmed euthanasia in federal legislation . . . when you kill people because they're in pain, that's euthanasia" (Bopp and Balch, 1985).

Judicial Application
of the Child Abuse Amendments

A county court in Minnesota relied on the Child Abuse Amendments in reviewing whether life-sustaining treatment should be withheld from an infant with disabilities. The infant, Lance Steinhaus, received severe brain damage from an abusive beating. The child's doctor and mother subsequently agreed that he should not receive antibiotics and should not be resuscitated. The Redwood County Welfare Department, which had custody of Lance since the injury, sought and obtained a temporary restraining order preventing termination of treatment. During two sets of hearings, the court received testimony that Lance was in a persistent vegetative state, and arguments were made that this condition fit the "chronically and irreversibly comatose" exception in which treatment need not be provided. The court found that "It's clear that the statute that's applicable does not make an exception to a persistent vegetative state so that the court would have to find that the child is in a chronic and irreversible coma in order for heroic measures beyond food, water and appropriate medication (t[o] be withheld)" (*In re Steinhaus*, Oct. 6, 1986; Transcript). The court relied heavily on the nonbinding guidelines issued by DHHS for guidance to the meaning of the Amendments (DHHS, 1983b). The court concluded that, because the Child Abuse Amendments require that all infants with disabilities receive "nutrition, hydration, and medication," antibiotics would have to be provided to Lance.

The Court stated:

> The Child Abuse Amendments of 1984 were enacted in the wake of considerable debate over whether children with disabilities should receive life-preserving treatment. The law is now clear that all infant children with life threatening conditions have a right to medically indicated treatment . . . in enacting the 1984 Amendments, the "quality of life" rationale was rejected (*In re Steinhaus*, Sept. 11, 1986).

Testimony by an independent medical expert at a second hearing convinced the court that Lance was not in a persistent vegetative state, but was in fact chronically and irreversibly comatose. Thus, the court held resuscitation was not required.

During the Steinhaus proceedings, the court clearly distinguished between patients in a comatose state and those in a persistent vegetative state. The court's careful following of the meaning of chronically and irreversibly comatose has since been questioned. Two physicians involved in the Steinhaus case, Stephen Smith and Ronald Cranford, neurologists at the Hennepin County Medical Center, have suggested that "permanently and completely unconscious" would be a better term to use as an exception to the Child Abuse Amendments, and David Coulter, a director of pediatric neurology, has stated his belief that the regulations actually refer to the persistent vegetative state, since children do not remain chronically and irreversibly comatose for long (Levine, 1986).

However, in the negotiations leading up to the consensus language formed in the Child Abuse Amendments, the term "unconscious" was explicitly rejected. "Unconscious" is defined as "insensible . . . not conscious" (*Dorland's*, 1988). In contrast, "coma" is defined as "[a] state of profound unconsciousness from which the patient cannot be aroused" (*Dorland's*, 1988). The term "coma" was, in fact, adopted because it was substantially more restrictive and was applicable to few children (Bopp and Balch, 1985). Thus, the consensus language was plainly intended to have narrow application, not the expansive application suggested by some.

Another case that has applied the Child Abuse Amendments of 1984 is *In re Baby Girl Muller* (Feb. 9, 1988). The infant, Rebecca Jean Muller, was born one month prematurely with severe, permanent mental and physical disabilities. She was unable to breathe without the assistance of a ventilator, unable to be fed except through intravenous means, and had damage to her brain, lungs, and kidneys (*In re Baby Girl Muller*, Feb. 9, 1988). She was described as "in a vegetative state . . . with minimal purposeful movements, and yet . . . responsive to pain" (*In re Baby Girl Muller* Feb. 9, 1988). Rebecca's parents requested the withdrawal of her ventilator, but requested the continuation of nutrition, hydration, and medication necessary for the comfort care of their daughter (*In re Baby Girl Muller*, Feb. 9, 1988). The guardian *ad litem* and attending physicians concurred with the parents' request.

Hospital officials and the parents petitioned a circuit court for a ruling on disconnecting life-support systems. How the Child Abuse Amendments of 1984 applied to the Muller case was discussed in a closed meeting attended by two doctors, the guardian *ad litem*, and attorneys representing the parents, the hospital, and the state. The guardian *ad litem* commented that all parties had the statute and "we're all trying to abide by the guidelines in it" (*Tampa Tribune*, Jan. 30, 1988). In order to comply with the Child Abuse Amendments guidelines, the Florida Department of Health and Rehabilitative Services (HRS) requested that an independent physician be appointed to examine the infant to assure that the court receive a complete report (*Tampa Tribune*, Jan. 30, 1988). By stipulation of the parties, an independent physician examined Rebecca. He determined that "use of the ventilator was merely prolonging (her) death and was not effective in ameliorating or correcting all of the infant's life threatening conditions" (*In re Baby Girl Muller*, Feb. 9, 1988). Based on the independent physician's conclusion, an HRS spokesman stated, "It's their opinion that the life-support equipment is only prolonging the infant's death. Therefore, HRS will not intervene" (*Tampa Tribune*, Feb. 4, 1988). Apparently, HRS determined through the independent examination that the infant's condition fit the second exception to the Child Abuse Amendments, which provides that treatment need not be provided if it would merely prolong dying.

The court found that Rebecca was "terminally ill, even with the aid of life-support systems" and agreed with the independent physician's report (*In re Baby Girl Muller*, Feb. 9, 1988). In light of the fact that the Child Abuse Amendments did not require treatment, the court held that the parents were "authorized to substitute their judgment for that of their infant daughter . . . to terminate the use of the ventilator . . . provided that all other medical care, including normal nutrition and hydration by intravenous means, if necessary, presently being given, continue" (*In re Baby Girl Muller*, Feb. 9, 1988). Rebecca Muller died the same day the ventilator was removed (*Tampa Tribune*, Feb. 10, 1988).

Criticisms of the
Child Abuse Amendments

One commentator has stated that the Child Abuse Amendments mandated changes in state law to adopt only reporting and referral procedures, but not the federal definition of medical neglect (Silver, 1989). In contrast, the regulations implementing the Child Abuse Amendments make clear that states, in order to qualify for federal funds for their child abuse programs, must meet the requirements of 45 CFR. 1340.15, which defines "medical neglect" (DHHS, 1983b). According to 45 CFR 1340.15, " 'medical neglect' includes, but is not limited to, the withholding of medically indicated treatment from a disabled infant with a life-threatening condition." The term "withholding of medically indicated treatment" means:

> The failure to respond to the infant's life-threatening conditions by providing treatment which [including appropriate nutrition, hydration, and medication], in the treating physician's [or physicians'] reasonable medical judgment, will be most likely to be effective in ameliorating or correcting all such conditions, except that the term does not include the failure to provide treatment [other than appropriate nutrition, hydration, or medication] to an infant when, in the treating physician's [or physicians'] reasonable medical judgment any of the following circumstances apply:

> (a) The infant is chronically and irreversibly comatose, (b) The provision of such treatment would merely prolong dying, not be effective in ameliorating or correcting all of the infant's life-threatening conditions, or otherwise be futile in terms of the survival of the infant, or (c) The provision of such treatment would be virtually futile in terms of the survival of the infant and the treatment itself under such circumstances would be inhumane" (DHHS, 1983b).

Thus, states are eligible for federal child abuse funds if they adopt the definition of medical neglect provided by the 1340.15. Other commentators have argued it is inappropriate for the government to mandate a standard of care ensuring survival of infants who will have disabilities because the government has not yet provided adequate rehabilitation and educational services for persons with disabilities (Newman, 1989). According to this line of reasoning, the government, in order to solve the problem of inadequate rehabilitation and education services for persons with disabilities, should eliminate the persons needing those services, thus eliminating the problem of inadequate services. Our society cannot tolerate a solution that proposes to rid society of the person rather than the problem.

State Compliance with the Child Abuse Amendments

The Child Abuse Amendments of 1984 give primary responsibility for enforcement to state Child Protective Services (CPS) agencies (DHHS, 1983b). In 1990, 47 states accepted federal funds for their child abuse and neglect programs, and therefore, agree to implement procedures to comply with the Child Abuse Amendments of 1984. Those states receiving federal funds for their child abuse and neglect programs must implement procedures to investigate and report suspected discriminatory denial of treatment (DHHS, 1983b).

However, a report issued by the United States Commission on Civil Rights has uncovered widespread and substantial failures of CPS agencies to carry out their statutory responsibilities (United States Commission on Civil Rights, 1989). Federal regulations require that state CPS agencies investigate suspected cases of medical neglect and determine whether treatment is medically indicated under the Child Abuse Amendments. However, according to the Commission Report, many state CPS agencies have abdicated their authority to internal hospital infant care review committees (ICRC), hospital staffs, and medical associations. Other CPS agencies have permitted the hospitals under investigation for medical neglect to name the "independent" medical authority who will rule on whether their course of treatment or nontreatment was proper (United States Commission on Civil Rights, 1989).

The Commission Report has shown that, in setting up procedures to investigate and respond to Baby Doe reports, CPS agencies have been more receptive to input from medical organizations—the groups representing those whose abuses the Child Abuse Amendments were enacted to curtail—and less receptive to input from disability organizations—the groups representing those whom the Amendments were designed to protect (United States Commission on Civil Rights, 1989). The Commission Report concluded: "The delegation of significant investigational responsibility by watchdog agencies to those they are supposed to be watching is perhaps the most serious form of widespread noncompliance by CPS agencies with the medical discrimination regulations it is their duty to implement" (United States Commission on Civil Rights, 1989).

Other ways in which CPS agencies fail to comply with the requirements of the Child Abuse Amendments of 1984 and their implementing regulations include having no written policies for securing the medical records, or for securing an independent medical examination of the infant with disabilities about whom a report of suspected medical neglect has been filed (United States Commission on Civil Rights, 1989). The Commission further found that CPS workers have failed to fulfill their obligations, because a majority of state CPS agencies are unclear in their policies concerning who is covered by the standards of treatment in the Act (United States Commission on Civil Rights, 1989). Moreover, CPS workers are as likely to use quality-of-life judgments as others who devalue the lives of persons with disabilities (United States Commission on Civil Rights, 1989).

A 1987 Department of Health and Human Services report revealed that 11 of 49 state CPS agencies took the position that "[b]ecause of medical and ethical issues involved, CPS responsibility for Baby Doe cases is not appropriate" (United States Commission on Civil Rights, 1989; DHHS, 1987). Ten other agencies were unwilling to state whether they regarded such responsibility as appropriate or not (United States Commission on Civil Rights, 1989; DHHS, 1987). The Commission Report found that these remarks and others, from state CPS personnel whose responsibility it is to investigate Baby Doe reports, reflected a grave weakness in the use of the Child Abuse Amendments to protect infants with disabilities. Overall, the Commission Report concluded that CPS agencies have failed in their responsibilities to enforce the Child Abuse Amendments effectively.

Failure of Infant Care Review Committees to Protect Infants with Disabilities

The Child Abuse Amendments of 1984 encouraged the establishment of Infant Care Review Committees, internal hospital committees that evaluate cases in which life-preserving medical treatment is being or may be withheld from infants with disabilities (Child Abuse Amendments, 1984). A 1986 survey found that 51.8% of hospitals with either a neonatal intensive care unit or over 1500 births annually had established such committees (University of Connecticut Health Center, 1987). However, ICRCs have been described as "very hit-and-miss" with regard to whether they effectively investigate cases (Protection of Handicapped Newborns, 1986).

Some commentators have promoted medical ethics committees rather than ICRCs, often designed less to enforce the treatment standards of the Child Abuse Amendments of 1984 than to protect doctors and hospitals from "intrusion" by the government or other outsiders (Strain, 1987; Shapiro and Barthel, 1986). "Ethics committees" are designed to provide a forum for the discussion of widely held opinions about withholding life-saving care from infants with disabilities. Decisions are made on a case-by-case basis varying from hospital to hospital, without consideration of a minimum level of care for these infants.

However, the ICRCs proposed by the Child Abuse Amendments of 1984 are based on model guidelines published by DHHS, which provide that "the basic policy [of ICRCs] should be to prevent the withholding of medically indicated treatment from disabled infants with life-threatening conditions" (DHHS, 1985b). DHHS intended ICRCs to serve as prognosis committees, requesting and evaluating specialized medical opinions to determine whether the facts in particular cases placed them under the Child Abuse Amendments. Thus, a reevaluation of the ethical and social factors about the propriety of the treatment is not warranted. Instead, an analysis of how the Child Abuse Amendments apply to the case should be discussed. The "ethical issues" concerning what circumstances justify withholding treatment should not be up for reconsideration on a case-by-case basis; they should be regarded as settled by the Child Abuse Amendments. Unfortunately, many hospitals have elected to take the "ethics committee" approach rather than that of Infant Care Review Committees, regardless of how they are named (US Commission on Civil Rights, 1989).

ICRCs have been found to be unwilling to follow the standard of care set forth by the Child Abuse Amendments. The committees often misdefine or more broadly define terms and exceptions, and incorporate consideration of the child's projected quality of life, a factor emphatically excluded from consideration under the Child Abuse Amendments (US Commission on Civil Rights, 1989).

Moreover, ICRCs are not fulfilling their legal obligations to report suspected denials of legally mandated treatment. The regulations require that ICRCs report to CPS agencies all suspected, not just known, cases in which life-saving medical care is being withheld (DHHS, 1983b). However, it seems that in only those cases in which there is disagreement among the physician, parents, and committee is a referral made to a CPS agency for investigation and possible legal action. Widespread failures to report are a serious concern, since many hospital ethics committees use a "nebulous and subjective" approach rather than applying the Child Abuse Amendments' standard of care (United States Commission on Civil Rights, 1989).

Finally, the disability perspective is often underrepresented on ICRCs (United States Commission on Civil Rights, 1989; Todres et al., 1988). (Physicians surveyed gave the least support for inclusion of a representative of a disability group on ICRC.) These committees are dominated by health-care personnel from the same hospital, who may see their primary function as protecting the hospital against potential legal liability for treating or not treating particular patients (Annas, 1984).

Independent scrutiny of ICRC proceedings, by medically knowledgeable and experienced disability advocates, and vigorously enforced reporting requirements are necessary for ICRCs to be effective in protecting infants with disabilities from being denied life-saving medical treatment (United States Commission on Civil Rights, 1989). Other factors that diminish the effectiveness of the Child Abuse Amendments of 1984 are discussed below.

Despite the protective definition of the "withholding of medically indicated treatment" contained in the Child Abuse Amendments, the practical utility of the statute as a basis for legal intervention to protect the lives of infants with disabilities is limited for several reasons. First, the statute is directed exclusively toward the operation of state child protective service systems receiving federal financial assistance from the Department of Health and Human Services (DHHS, 1983b). Only state child abuse agencies are authorized to act to protect infants with disabilities in danger of medical neglect. As a result, enforcement of the substantive guarantees of the statute (i.e., protection against medical neglect) is left entirely to the states with federal health or civil rights officials unable to intervene directly in a Baby Doe crisis, even where state action or inaction is clearly unlawful.

Second, the statute creates no private right of action for parents or other interested persons to bring a civil action in federal court to prevent the unlawful withholding of medically indicated treatment from a handicapped infant, and it is unlikely that "outside" parent or disability groups would have standing to sue the state child protective service system to enforce the provisions of the Act. Also, the parents, who are often victims of discriminatory practices by physicians, are not authorized to bring an action to protect their child.

Third, federal enforcement options are limited and cumbersome. It would appear that only two enforcement approaches are available. The Department of Health and Human Services could initiate administrative proceedings to terminate the child abuse grant fund, alleging that the state child protective service agency had violated the conditions under which the grant was made. In the alternative, it appears that the secretary of Health and Human Services could request that the attorney general initiate a civil action to enforce the grant conditions through injunctive relief. (The federal government has the right to sue for enforcement of its contractual rights, including the conditions of grant awards) (*Rex Trailer Co. v. United States*, 1955; *United States v. Marion County School District,* 1980). This second alternative is particularly important, given the current public confusion surrounding the issue and the history of hostility by many child protective service workers to being drawn into what they believe is essentially a "political" issue. It is not inconceivable, in this climate, that a state might turn down federal child abuse funds rather than comply with the new rule.

Fourth, the Child Abuse Amendments deal with the denial of treatment only on a case-by-case basis, essentially making the Act ineffective against an institutionally adopted pattern and practice of discrimination. Finally, if an infant is dead or so debilitated by the very medical neglect in question that medical treatment would no longer be beneficial, the Child Abuse Amendments provide no remedy for the infant, only termination of federal funds to the child abuse and neglect agency.

Conclusion

The Child Abuse Amendments were enacted to answer the call for protection of disabled infants with life-threatening conditions from discriminatory denial of medical treatment. The language embodied by the Child Abuse Amendments was carefully developed by a consensus of groups with diverse interests in this issue. These regulations provide, for the first time, a uniform minimum standard of care for all disabled infants with life-threatening conditions, so that their care may no longer be subject to the biases of their caregivers. Despite criticisms and suggested interpretations, the Child Abuse Amendments provide a clear standard of care. However, weaknesses of the Child Abuse Amendments lie in their enforcement. Many state CPS agencies, entrusted with the responsibility of investigating and dealing with suspected and known cases of medical neglect, have failed to comply with their statutory obligations. Moreover, some ICRCs, which were proposed for the purpose of applying the law to particular cases, have taken a case-by-case subjective decision-making approach while ignoring the standard of the law. Despite these weaknesses and criticisms, the Child Abuse Amendments, if applied properly, are an effective protective mechanism for disabled infants with life-threatening conditions.

References

American Academy of Pediatrics v. Heckler, 561 F. Supp. 395 (D.D.C. 1983).

American Hospital Association v. Heckler, 585 F. Supp. 541 (S.D.N.Y. 1984).

Annas, G. (1984) Ethics committees in neonatal care: Substantive protection or procedural diversion? *Am. J. Public Health* **74,** 843–845.

Barber v. People, 147 Cal. App. 2d 1006, 195 Cal. Rptr. 484 (1983).

Bopp, J. (1985) Legal implications of medical procedures effecting the born, in *Human Life and Health Care Ethics* (University Publication of America, Frederick, MD), pp. 284–287.

Bopp, J. and Balch, T. (1985) The Child Abuse Amendments of 1984 and their implementing regulations: A summary. *Issues Law Med.* **1**, 91–130.

Bowen v. American Hospital Association, 476 US 610 (1986).

Child Abuse Amendments of 1984. 42 U.S.C.A. 5101–5103 (1984).

Child Abuse Prevention Challenge Grants Reauthorization Act of 1989. 42 U.S.C.A. 5101–5117 (Supp. 1989).

Congressional Record (1984a) 130:21,156.

Congressional Record (1984b) 130:21,153 (Joint Explanatory Statement by Principal Sponsors of Compromise Amendments Regarding Services and Treatment for Disabled Infants).

Derr, P. (1986) Nutrition and hydration as elective therapy. *Issues Law Med.* **2**, 25–38

DHHS (1983a) Nondiscrimination on the Basis of Handicap. Interim Final Rule. 48 Fed. Reg. 9630.

DHHS (1983b) Child Abuse and Neglect Prevention and Treatment Program. 45 Code Fed. Reg. 1340. 14, .15(b)(2)-(3)(ii), (c), Appendix.

DHHS (1984a) Nondiscrimination on the Basis of Handicap, Procedures and Guidelines Relating to Health Care of Handicapped Infants. Final Rule 49 Fed. Reg. 1622 (codified at 45 CFR 84.55).

DHHS (1984b) Child Abuse and Neglect Prevention and Treatments Programs. 49 Fed. Reg. 48,160.

DHHS (1985a) Child Abuse and Neglect Prevention and Treatments Programs, Final Rules. 50 Fed. Reg. 14,878.

DHHS (1985b) Services and Treatment for Disabled Infants; Model Guidelines for Health Care Providers to Establish Infant Care Review Committees. 50 Fed. Reg. 14,893.

DHHS (1987) Office of Inspector General. Survey of Baby Doe Programs. GPO, Washington, DC.

Dorland's Illustrated Medical Dictionary (1988) (W. B. Saunders, Philadelphia), pp. 363, 1785.

Final Rules. 50 Fed. Reg. 14,878.

Gerry, M. and Nimz, M. (1987) Federal role in protecting Babies Doe. *Issues Law Med.* **2**, 339–377.

House Conference Report (1984) No. 1038, 98th Congress, 2d Session.

In re Baby Girl Muller, No. 88-1073 (Hillsborough County Cir. Ct. Feb. 9, 1988).

In re Conroy, 98 NJ 321, 486 A.2d 1209 (1985).

In re Steinhaus, No. J-86-92 (Minn. Redwood County Ct., Juv. Ct. Div. Sept. 11, 1986) (Transcript, Oct. 6, 1986).

In re Steinhaus, No. J-86-92 (Minn. Redwood County Ct., Juv. Ct. Div. Sept. 11, 1986), reprinted in *Issues Law Med.* **2**, 241–252.

Levine, C. (1986) Confusion over the language of the Baby Doe regulations. *Hastings Cent. Rep.* **16,** 2–4.

Lynn, J. and Childress, J. F. (1983) Must patients always be given food and water? *Hastings Cent. Rep.* **13,** 17–21.

Nerney, T. (1987) Telephone interview between Thomas Nerney, a disability rights advocate involved in Child Abuse Amendments negotiations, and coauthor Mary Nimz.

Newman, S. A. (1989) Baby Doe, Congress and the states: Challenging the Federal Treatment Standard for impaired infants. *Am. J. Law Med.* **15,** 1–60.

Paris, J. J. and Fletcher, A. B. (1983) Infant Doe regulations and the absolute requirement to use nourishment and fluids for the dying infant. *Law Med. Health Care* **11,** 210–213.

Protection of Handicapped Newborns (1986) Hearing Before the United States Commission on Civil Rights vol. II, pp. 35, 36.

Rex Trailer Co. v. United States, 350 US 148 76 S. Ct. 219 (1955).

Senate Bill 1003 (April 1984a) Proposed Amendments from Senator Edward Kennedy's Staff.

Senate Bill 1003 (April 1984b) Proposed Amendments from the American Academy of Pediatrics.

Senate Bill 1003 (April 1984c) Second Draft of Proposed Amendments from Senator Edward Kennedy's Staff.

Senate Bill 1003 (May 18, 1984) Senate Labor and Human Resources Staff Discussion Draft of Proposed Amendments.

Senate Bill 1003 (May 22, 1984) Areas of Concern Regarding 5/18/84 Discussion Draft of Proposed Amendments.

Senate Bill 1003 (June 7, 1984) Second Senate Labor and Human Resources Staff Discussion Draft of Proposed Amendments.

Senate Bill 1003 (June 13, 1984) Memorandum Regarding Proposed Amendments to Senate Labor and Human Resources Staff from Jan Carroll, National Right to Life Committee, Richard Doerflinger, NCCB Committee for Pro-Life Activities, and, Doug Badger, Christian Action Council.

Senate Bill 1003 (June 14, 1984) Memorandum re: Staff Response to Proposed Amendments to June 7 Discussion Draft of Amendments from Jan Carroll.

Senate Bill 1003 (June 19, 1984) Memorandum to Staff/Organizations from Susanne Martinez.

Senate Bill 1003 (June 25, 1984) Proposed Amendment by Medical Groups.

Senate Report (1974) No. 1297. 93rd Congress, 2d Session, p. 38.

Shapiro, R. and Barthel, R. (1986) Infant Care Review Committees: An effective approach to the Baby Doe dilemma? *Hastings Law J.* **37,** 827–862.

Silver, J. D. (1989) Baby Doe: The incomplete federal response. *Family Law Q.* **20,** 173–185.

Strain, P. (1987) Bloomington to Here: A Brief History of the Baby Doe Regulation, Medical Neglect and the Disabled Infant: The Impact of the Baby Doe Regulation.

Tampa Tribune. Jan. 30, 1988. 1B, col. 5.

Tampa Tribune. Feb. 4, 1988. 12B, col. 1.

Tampa Tribune. Feb. 10, 1988. 1B, col. 1.

Todres, I., Guillemin, J., Grodin, M., and Batten, D. (1988) Life-saving therapy for newborns: A questionnaire survey in the state of Massachusetts. *Pediatrics* **81,** 643–649.

United States Commission on Civil Rights (1989) Medical Discrimination Against Children with Disabilities, pp. 111–117, 127–129.

United States v. Marion County School District, 625 F.2d 607 (5th Cir. 1980).

United States v. University Hospital, State University of New York at Stony Brook, 575 F. Supp. 607 (E.D.N.Y. 1983).

United States v. University Hospital, State University of New York at Stony Brook 729 F.2d 144 (2d Cir. 1984).

University of Connecticut Health Center (1987) Pediatric Research and Training Center, National Collaborative Survey of Infant Care Review Committees in United States Hospitals, p. 6.

Hard Cases
Make Bad Law

The Legacy of the Baby Doe Controversy

Arthur L. Caplan

Many years have passed since the heated controversy flared over the wisdom of the federal government enacting the so-called "Baby Doe" regulations for the care of newborns. These regulations, originally promulgated in March 1983 by the Civil Rights Division of the Department of Health and Human Services (DHHS), required the posting of warning notices in neonatal nurseries concerning the duty to treat handicapped newborns and called for the creation of federal "flying squads" of medical investigators to rush to hospitals where discrimination in treatment had been alleged (DHHS, 1983). These regulations, or more accurately, a modified version of them issued on January 12, 1984, were eventually held to be unconstitutional by the United States Supreme Court in a 1986 decision, *Bowen v. Am. Hosp. Ass'n.*

In October 1984, Congress, after lengthy negotiations with provider organizations, professional societies, disability groups, and right-to-life groups, enacted the Child Abuse Amendments of 1984. This law led to federal regulations being issued in April of 1985 governing the care of newborns

From: *Compelled Compassion* Eds.: Caplan, Blank, and Merrick
©1992 The Humana Press Inc.

(DHHS, 1985). These regulations require the treatment of all newborns except:

1. When the infant is chronically and irreversibly comatose;
2. When the treatment would merely prolong dying;
3. When treatment would not be effective in ameliorating or correcting all of an infant's life-threatening conditions;
4. When treatment would otherwise be futile; or
5. When the provision of treatment would be virtually futile and would be inhumane.

Under all circumstances, nutrition, hydration, and medication are to be provided to infants. State child protective services are given the responsibility of monitoring and enforcing these regulations.

Analyzing the impact of the debate about the Baby Doe regulations and the Child Abuse Amendments of 1984 on infants, their parents, and health-care providers is no simple task. Medical care for newborns has hardly remained static in the years since the first Baby Doe regulations were promulgated. Professional and public opinion concerning the medical treatment of newborns, the acceptability of discontinuing medical treatment for any patient, and the significance of impairment in influencing the quality of life a child or an adult may enjoy have changed since the passage of the Child Abuse Amendments.

At the time of the controversy over the need for federal regulations, many who were critical of the effort to permit governmental oversight of clinical decision making (Murray and Caplan, 1985; Rhoden, 1986; Caplan, 1984, 1987) argued that the Baby Doe regulations were superfluous. They solved a problem that was limited in scope, and had no clear-cut right answers with methods that were inappropriate and even offensively intrusive in clinical settings. Those who believed federal oversight was appropriate and long overdue argued that the federal government had a legitimate interest in protecting the lives of newborn children and not only should, but must, intervene to protect their rights (DHSS, 1983; Gerry, 1985; Wilkie, 1991).

Why was there such a disparity of opinion about the treatment of newborn infants in American hospitals? Why did persons concerned about the best interests of children come to such disparate conclusions about the proper role for government, parents, and third parties with respect to their medical care? Have the laws and regulations that resulted from the Baby Doe controversies had their intended result? To answer these questions, it is necessary to understand the kinds of issues and the sorts of value disagreements that divided the parties in the controversy.

The Cases that Triggered the Controversy

The initial national debate about the medical treatment of newborns was triggered by the famous Baby Doe case in which a baby, born on April 9, 1982 in Bloomington, Indiana, with Downs syndrome, esophageal atresia, and tracheo-esophageal fistula, died six days later subsequent to his parents' refusal to consent to potentially life-saving surgery. The controversy deepened with the birth of Baby Jane Doe on October 11, 1983 on Long Island, New York. This child was born with spina bifida, hydrocephalus, microcephaly, spasticity, and an apparently malformed brainstem. Her parents, given the option of a surgical intervention to help treat the hydrocephalus or a conservative course of nursing care, chose not to consent to the surgery (Murray and Caplan, 1985).

In both cases, efforts were made by third parties to obtain intervention by courts in order to assure access to medically efficacious treatment for these infants. In the Baby Doe case, officials from the state of Indiana were in the process of seeking an order from a US Supreme Court Justice to compel treatment when the newborn died. In the Baby Jane Doe case, the federal government attempted to intervene to obtain the medical records of the child in order to ascertain whether treatment would be beneficial. A storm of controversy ensued over the appropriateness of any third-party interventions in cases such as these (Weir, 1984; Murray and Caplan, 1985).

It is interesting that the debate in the United Kingdom over the nontreatment of newborn infants during the early 1980s also focused on cases involving newborns with congenital impairments. In the 1981 case known as "re B (Minor)," the parents of a child with Downs syndrome and duodenal atresia decided not to have surgery performed to correct the atresia, knowing this would mean the death of their baby. Most of the doctors they consulted agreed with their position, but not all did. Eventually, legal proceedings commenced that resulted in the baby being made a ward of the court. The baby was operated upon and placed in foster care.

The other paradigmatic British case involved the decision of a doctor to abide by parental wishes not to provide food or water to a child with Downs syndrome. The pediatrician, Leonard Arthur, was originally charged with murder and later tried for attempted murder. He was acquitted after a widely publicized and controversial trial (Brahams and Brahams, 1983).

These cases became the paradigms or exemplars that grounded public and professional debate about the treatment of newborns in the English speaking world.

Did and Does the Federal Government Have a Place at the Bedside?

Some critics of the Baby Doe regulations maintained that there were simply no circumstances that would ever justify outside intervention by federal officials into the practices of neonatal nurseries. Local and state government, child protective services, as well as local and state courts, were seen as entirely adequate to the task of regulating the clinical care given to newborn children. Even simple requests by federal officials to review records and medical histories of particular children were criticized as being unduly intrusive.

The dispute about the proper role of the federal government in assuring the civil rights of children was fraught with irony. Many who had argued for an aggressive federal role in enforcing civil rights laws with respect to minorities and women, such as the ACLU, professional associations of lawyers, doctors, and nurses, and feminists maintained that the federal government had no business snooping around neonatal units. Many conservative groups, public officials, and civic organizations who on other issues maintained that the federal government should play a minimal role in the lives of citizens, and who touted the glories of local control and individual autonomy in resolving matters of individual behavior were in the front ranks of the effort to assign an oversight and enforcement role to the federal government with respect to treatment decisions for newborns.

For example, the same civil rights division officials in the Reagan administration who pressed the case in the federal courts for recognizing the legitimacy and appropriateness of the original set of Baby Doe regulations also argued, at roughly the same time, in the Grove City College case that the federal government had no right to impose federal civil rights statutes upon Grove City College. They maintained in federal courts that colleges that accepted federal money to support particular programs were not bound to apply federal civil rights laws governing women and minorities to other programs not directly and explicitly supported by federal funds. Yet, these same officials argued that Section 504 of the Rehabilitation Act of 1973 did apply to the care of newborns in any American hospital, public or private, on the grounds that hospitals that accepted federal money, including Medicare and Medicaid payments, were bound by the law in all of their programs, even though the 1973 law made no explicit mention of newborns or medical care.

The disagreement about the need to monitor and regulate the treatment of newborns was in many ways a stalking horse for a more general debate about the proper role of government with respect to clinical autonomy in health care. In fact, much of the resistance on the part of organized medicine, especially the American Medical Association and various state medical societies who weighed in in opposition to the efforts of the administration and some members of Congress to legislate and regulate in the area of the medical care of newborns, stemmed as much from a general resistance to any sort of government intrusion into the practice of medicine as it did a specific opposition to the presence of federal officials in neonatal nurseries or to the rights of parents to decide not to pursue medical care for their infants.

Although some believe (Nimz, 1989) that the job of protecting the welfare of newborns belongs to the federal government, since state child protective services have bungled the job or *been* unwilling to take the job seriously, this view has relatively few adherents. Legally, the monitoring of abuse and neglect has long been a responsibility of the states, not the federal government. Indeed, there is a certain amount of variation permitted to the states with respect to what constitutes neglect or abuse in such areas as religious refusals of treatment.

The reality of newborn care is that, unless a parent, a physician, a nurse, or a social worker believes something is amiss in the care of a child and reports it, there is no way in which government can effectively monitor the medical care newborns receive. Even accrediting organizations in health care must rely upon the good-faith quality control of providers and families in order to know whether the care given by a particular hospital is adequate. Since this is so, the case for transferring oversight responsibility to the federal government or a federal agency has not yet been made. There are too many births, events move too quickly, and the system is too reliant on the monitoring and whistle-blowing activities of providers and parents for any federal agency to effectively monitor or intervene.

This does not mean that the federal government and federal agencies ought not have some ability to review cases and retrospectively assess care. It is one thing to allow federal officials outside the clinical setting the right to override provider decisions in the neonatal nursery. It is a different matter to argue that federal agencies ought to have the authority to see charts and records retrospectively in order to assess the quality of care being provided in neonatal units. The latter function is consistent with the federal government's responsibility to assure that minimal standards quality of medical care are delivered at institutions where the federal government is directly paying for the care of patients, including newborns.

Babies Doe and the Scope of Law and Regulations

Some of the most vehement critics of a federal role in monitoring the care of newborns felt that the content of the proposed federal regulations and, later, law was seriously flawed, because the paradigmatic cases used in articulating the scope and meaning of the regulations were flawed (Caplan, 1984; Murray and Caplan, 1985). The cases that motivated concern on the part of the Surgeon General, the Department of Health and Human Services, and ultimately, many members of Congress and the President about the adequacy of the care being given to newborn children in neonatal units were cases that focused on the care of children with Downs syndrome or spina bifida. This focus, although appropriate for those who felt the infants in the Baby Doe and Baby Jane Doe cases did not receive proper medical care, created problems for the formulation of public policy.

First, treatment practices with respect to spina bifida and Downs syndrome were rapidly changing from what they had been in the 1960s and 1970s. Federal officials, right-to-life groups, and some disability groups were seeking changes in medical practices that were changing of their own momentum.

Second, infants with these impairments did not adequately represent the range or complexity of treatment decisions associated with other kinds of anomalies, impairments, and disabilities in newborn infants. The kinds of cases and treatment issues addressed by existing federal law and regulations under the Child Abuse Amendments of 1984 were not in the early 1980s, and are certainly not in the early 1990s, the only ones that raised difficult moral problems for parents, health-care providers, and society.

Current regulations and existing law address the issue of the treatment of newborns under the framework of child abuse and neglect, but the original attempt to regulate the treatment of newborns was promoted under a framework of discrimination against the handicapped. The legal foundation upon which the original Baby Doe regulations rested was the prohibition against discrimination against those with disabilities as enjoined by Section 504 of the Rehabilitation Act of 1973. The discrimination focus was a result of the fact that much of the concern about the fate of children in neonatal nurseries was exclusively focused on infants born with congenital impairments and disabilities.

The scope of later versions of the regulations, and eventually the Child Abuse Amendments of 1984, expanded to include discussions of when treatment could and could not be withheld from children born very prematurely, with very low birthweight, or with injuries resulting from drugs, abuse, or the process of birth itself. After the Supreme Court invalidated the appropriateness of the discrimination framework, the foundation of the law changed to abuse and neglect (Rhoden and Arras, 1985). The range of cases captured under this framework involves many different kinds of treatment decisions for a very diverse set of newborns, some with impairments, some whose only "impairment" was extreme prematurity, and some with impairments so severe that they were incompatible with life regardless of the medical care that might be given. It is simply wrong to equate the treatment decisions that parents and providers face in deciding what to do about the fate of a 450-gram neonate with the choices that are faced when a baby with Downs syndrome requires surgery to repair a hole in his stomach. However, despite the shift in the philosophical foundation for federal regulation, the change in treatment practices for spina bifida and Downs syndrome, and the expansion of the range of cases to be covered, treatment decisions about infants born with Downs syndrome or spina bifida dominated and, to some extent, continue to dominate both political and popular discussions of the appropriateness of legislating medical care by the federal government.

The Impact of Federal Law and Regulation: Have the Baby Doe Regulations Done Any Good?

Relatively few studies have been conducted of the impact of federal law and regulations, and in some parts of the country, state laws, on the practice of neonatology or the treatment of newborns. The information that does exist does not support the view that federal intervention was necessary, nor does it support the position that federal action has been beneficial for infants or their families.

In 1987, the Office of the Inspector General of the Department of Health and Human Services conducted a survey of state child abuse and neglect agencies to determine compliance with the 1984 federal law. The investigation found that, in the two years since the regulations had been issued, there were a total of 22 reports of the abuse or neglect of infants under Baby Doe regulations. State child protective agencies believed that the evidence was such as to merit an investigation in six cases, an astoundingly miniscule percentage of the possible pool of cases, since there were more than 3,000,000 births in American hospitals during this period. It was not clear whether state agency intervention had changed the treatment of a newborn in any of the six cases.

Reports of infant abuse or neglect under the child abuse regulations binding child protective services departments, in many states (a few states have declined federal funds for child abuse and are, as a result, not bound by the regulations), have, if anything, decreased since 1987. The fact that no physician, hospital, or nurse in the United States has been found civilly or criminally liable under the provisions of the 1984 act and 1985 regulations would tend to corroborate the position that the law and regulations now on the books were not merited by a "holocaust" in American neonatal units, nor has federal intervention under the framework of child abuse and neglect helped shed much light on the key moral dilemmas that still trouble parents and their providers today in caring for neonates (Fost, this volume).

The few subsequent studies of neonatal practices and provider attitudes that have been done show that some newborns may have received overly aggressive and arguably inappropriate care (Kopelman et al., 1988; Young and Stevenson, 1990). They hint that federal law and regulation may be doing more harm than good for infants and their families by forcing interventions upon infants that are simply not justified in terms of their efficacy or by allowing some providers to fuzz the line between research and therapy.

Ironically, the major outcome of the Baby Doe regulations seems to be that some infants who are born extremely premature wind up getting full-press, aggressive interventions with less choice being given to their parents. Experimentation and innovation in neonatology have flourished with respect to premature infants. The medical care of children born with mild or moderate congenital impairments seems to have remained stable throughout the late 1980s. There has undoubtedly been much more discussion than would have been the case had the Baby Doe controversy never erupted about the medical care and rights of children with disabilities. However, ironically, the level and degree of care given to children with spina bifida or Downs syndrome today seem as much a function of increased understanding of these conditions by health-care providers as they do federal efforts to prohibit or penalize discrimination.

Paradigmatic Cases
and the Meaning of Disability

Recent scholarship in bioethics has emphasized the importance of paradigmatic cases in understanding how moral analysis produces consensus (Caplan, 1989; Hoffmeister et al., 1989). Nowhere is the role of paradigmatic cases more in evidence than in the Baby Doe controversy. The emergence of the Bloomington Baby Doe and the Long Island Baby Jane Doe cases as defining the scope and content of moral debate for most Americans meant that the focus of this debate was on children whose needs and problems did not, and still do not, represent the full range of newborn children whose medical problems raise complex ethical questions about starting and stopping treatment (Weir and Bale, 1989; Bay and Burgess, 1991).

The battle over the role government should take in protecting the welfare and health of newborns was fought in the early 1980s on the terrain of disability and handicap, and within that domain, only a narrow set of impairments were used as exemplars. The Baby Doe, Baby B (minor), and Baby Jane Doe cases involved decisions by parents not to offer medical treatment to children who were born with congenital impairments. The presence of impairment in these cases raised the question of to what extent decisions were being made by parents, physicians, or both to allow newborn children to die or go without possibly beneficial treatment for the sole reason that the children had been born with impairments.

There can be no doubt that some infants born with congenital impairments were not treated by doctors in the hope that they would die and some of them did die (Duff and Campbell, 1973; Lorber, 1972; Gerry, 1985). There should be no doubting the fact that it was then and remains appropriate now for third parties to inquire about and comment upon the medical care given to newborn infants with impairments.

However, the nontreatment of newborns with congenital disabilities, such as Downs syndrome or spina bifida, seems to have been a phenomenon much more characteristic of the 1960s and 1970s than of the 1980s.

One of the most frequently cited articles by those who believed that the "holocaust" in American neonatal units demanded prompt and decisive action by the federal government was a piece in the *New England Journal of Medicine* by Raymond Duff and A. G. M. Campbell that reported that at the Yale–New Haven hospital, from 1970 to 1972, 43 of the 299 deaths that occurred in the special care nursery followed decisions to withdraw treatment (Duff and Campbell, 1973). There should be no doubt that many infants with spina bifida, Downs syndrome, and other congenital disorders were allowed to die in neonatal nurseries in the 1960s and early 1970s in the United States and in the United Kingdom. However, the prevention of deaths, such as those that occurred in the 1960s and early 1970s, was, in the eyes of many health-care professionals, a moot point by the early 1980s. In the 1960s, interventions such as artificial feeding and fluid provision were not well understood for infants. Even the efficacy of shunts in the treatment of hydrocephalus remained controversial (Lorber, 1972). This was no longer true by the mid1980s, when the use of shunts and the management of technology in intensive care units had made great strides.

A major impact of the federal law and regulations that were enacted as a result of the controversy (and it has been an important and positive one) was to affirm the rights, worth, and interests of children born with impairments and congenital birth defects. However, the fight over the DHHS regulations and continuing disagreements about the impact of and need for subsequent law were greatly complicated by the fact that the parties involved in the debate did not see eye to eye about the need for an affirmation of the rights and worth of newborns with congenital impairments.

The overwhelming majority of health-care professionals were, by the time the controversy erupted, convinced that there was much they could do to treat infants and help the families of infants born with congenital anomalies. Those in the disability and right-to-life movements were not convinced that the interests of these children would be adequately protected by health-care providers and families who might still have strong biases against disability (Asch and Fine, 1988; Gerry, 1985). Health-care professionals involved in the care of newborns were much more troubled in the early 1980s about the treatment of newborns who were born extremely premature, or with severe illnesses or injuries. They believed that the Baby Doe laws and regulations were of little help, in that they applied standards appropriate for one category of newborn, those with congenital impairment, to other children whose medical care was seen as very different (Bay and Burgess, 1991). The problem of "apples and oranges" with respect to which infants were the paradigmatic source of moral problems in the neonatal nursery in the early 1980s was a major source of confusion and controversy in the debate over the original Baby Doe rules.

Part of the problem with focusing only upon cases of spina bifida and Downs syndrome as paradigmatic is that other sorts of disabilities, far more severe and devastating, were equated with these. For example, a child born with anencephaly might be described as disabled, but it would seem more accurate to describe a child with most of its brain missing as unabled rather than disabled. Similarly, to call a child with Lesch-Nyhan's Syndrome or Tay-Sachs disabled is to lump together under the disability rubric conditions that connote far more than the impairment of capacities and abilities generally enjoyed by most of the population.

On the other hand, focusing on only certain forms of disability obscures the fact that many children born in neonatal units are not disabled, but are at risk of disability because of prematurity. A child born at 450 grams is not accurately described as disabled. That infant may develop disabilities or may not, depending both upon the care given and a variety of little-understood factors. To say that medical treatment must be given to a 450-gram neonate on the grounds that to deny care is to neglect an infant because it is disabled or may become disabled is simply to waffle the distinction between experimentation and treatment. No one knows what will happen if treatment is given to babies born extremely prematurely. When outcomes are unknown and highly uncertain, it seems conceptually muddled to lump the treatment of children in this category with those who are much older, but born with congenital impairments.

Do Parents Care About Their Kids?

In the time I have spent observing neonatal nurseries, nothing has impressed me more than the fact that the overwhelming majority of parents love their newborn infants and want to do what is best for them. No matter how sick, how impaired, or how different the babies are, nearly all parents I have seen have quickly bonded to their newborn babies and have sought only the best of care for their children. This bond between parent and child has been reported by many others (parents and social scientists) who have spent time around newborn nurseries (Guillemin and Holmstrom, 1987). There certainly are parents who would abandon or even injure a child whom they perceived to be less than perfect or burdensome (Asch and Fine, 1988; Murray and Caplan, 1985; Powell and Hecimovic, 1985), but such parents are in the clear and unequivocal minority.

There is absolutely no reason to think that parents in the early 1970s were any less likely to bond to and love their babies than was true in the 1980s or today. The difference is that, in the 1960s, Downs syndrome and spina bifida were both considered conditions that would lead to inevitable lives of misery for those with the conditions (Crocker and Cullinane, 1972; Lorber, 1972). Parents were told that no cures were possible, that severe retardation and disability were inevitable, and that the ultimate and sole destination for most children with Downs syndrome or spina bifida would be a state-run institution or asylum—an institution that parents often believed to be underfunded and horrible. With this kind of bleak prognosis, it is not surprising that many parents would elect nontreatment for their children. For many parents, it must have seemed to them kinder to allow their child to die than to face a life of inevitable misery (Powell and Hecimovic, 1985). However, a decision for nontreatment cannot be equated, as those who refuse medical treatment for their children on religious grounds correctly point out, with indifference or hostility toward one's own child.

In the two decades that have elapsed since the publication of Duff and Campbell's paper and the publication of Lorber about his experience in deciding whether or not to treat newborns with spina bifida in his nursery, much has changed in our knowledge, understanding, and ability to treat congenital anomalies. Experts now know, having been pushed to think more clearly about handicaps by disability groups and parent associations, that Downs syndrome refers to a range of potential outcomes, not a single, stereotypic presentation. The same is true for spina bifida. It is also known that early intervention and training can get results in allowing children with these conditions to maximize their capacities and abilities. Also, far more is known about treatments and coping strategies for congenital disorders than was available at Yale or Sheffield in 1970.

Looking through the retrospectoscope, it is difficult to understand a decision to allow a child born with spina bifida to remain unshunted and untreated. Yet, medical opinion concerning spina bifida and Downs syndrome was so dire in the 1960s and 1970s that recommendations not to treat were commonplace.

One of the first cases presented to me when I was teaching at Columbia University's Medical Center, in the late 1970s, was that of a newborn infant, Ellen, who was born with spina bifida and a variety of other anomalies. Her doctors were convinced that the lesion was so high on the spinal cord that the baby would always be paralyzed from the waist down, and incontinent of both bowel and bladder. They also felt the degree of hydrocephalus already present had so damaged the baby's brain that she would be permanently and severely retarded. They advised the family against any treatment, including the placement of a shunt. A number of them told me that they hoped Ellen would succumb to an infection, since her prognosis was grim indeed. I was persuaded by their predictions of a life of utter suffering and misery that the right thing to do was not to treat Ellen.

Ellen's parents were not. The family refused to listen to these grim forecasts about their daughter. They had no previous children, and Ellen was very much a wanted child. They insisted she be shunted. They took her home and gave her meticulous nursing care. The father later spent much of his time building toys and special transportation equipment for his paralyzed daughter.

Ellen, by the age of six, had become a vivacious, charming, and outgoing kid. She was paralyzed from the waist down and confined to a wheelchair, she was incontinent, and she required constant help in feeding, bathing, toileting, and other activities of daily living, but Ellen was certainly much loved and giving a lot of love back to her parents. The prospect that she might have been allowed to die filled them with nothing but dread. The fact that I had concurred with those, six years earlier, who said she would be better off dead left me with the same feeling.

Today, many argue that the same outcome might have come true for Baby Jane Doe if her parents had been cajoled or forced into having her shunted sooner than they did by the intervention of the federal government (Andrusko, 1991; Koop, 1991; Wilkie, 1991), and they may be correct. However, the subsequent demonstration that at least some infants proved their doctors wrong does not prove either that all did, or that the physicians and nurses who believed this prognosis for infants with spina bifida or Downs syndrome were not acting in good faith or even that they were wrong. It does show that medical decision making sometimes must proceed with a high degree of uncertainty, and that different parents and doctors may respond differently in the face of uncertainty (Rhoden, 1986; Caplan, 1987). Whether it was right or is now right to attempt to legislate a single standard of care in response to uncertainty is much debated. My own view is that it was not then and is not now appropriate for the federal government to try to be the agent that has responsibility for resolving clinical uncertainty. Public policy must presume that parents love their children, seek what is best for them, and act accordingly. The burden should be on others, be they providers or government officials, to show that these presumptions are false. That is not the spirit that motivated the original Baby Doe regulations and, unfortunately, it is not always the spirit that guides their application in the neonatal and pediatric intensive care units of today.

References

Andrusko, D. (1991) *NRL News.*

Asch, A. and Fine, M., eds. (1988) Moving disability beyond stigma. *J. Soc. Issues* **44,** 1.

Bay, B. and Burgess, M. (1991) A survey of Calgary pediatricians' attitudes regarding the treatment of defective newborns. *Bioethics* **5(2),** 139–149.

Brahams, D. and Brahams, M. (1983) The Arthur case—A proposal for legislation. *J. Med. Ethics* **9,** 12–17.

Caplan, A. (1984) Is it a life? *The Nation* **238(2),** 37.

Caplan, A. (1987) Imperiled newborns. *Hastings Cent. Rep.* **17:6,** 15–32.

Caplan, A. (1989) Moral experts and moral expertise. in*The Nature of Clinical Ethics,* B. Hoffmaster, B. Freedman and G. Fraser, Eds., Humana Press, Totowa, NJ, 59–88.

Crocker, A. and Cullinane, M. (1972) Families under stress. *Postgrad. Med.* **21,** 223–229.

Department of Health and Human Services (DHHS) (1983) Nondiscrimination on the basis of handicap. 48 *Fed. Reg.,* 9630–9632.

Department of Health and Human Services (DHHS) (1985) Child abuse and neglect prevention and treatment program, and services and treatment for disabled infants: Model guidelines for health care providers to establish Infant Care Review Committees. 50 *Fed. Reg.,* 14878–14901.

Duff, R. and Campbell, A. (1973) Moral and ethical dilemmas in the special-care nursery. *N. Engl. J. Med.* **289,** 890–894.

Fost, N. (1992) Infant Care Review Committees, this vol.

Gerry, M. (1985) The civil rights of handicapped infants. *Issues Law Med.* **1,** 15–66.

Guillemin, J. and Holmstrom, L. (1987) *Mixed Blessings* (Oxford, New York).

Hoffmeister, B., Freedman, B., and Fraser, G., eds. (1989) *Clinical Ethics: Theory and Practice* (Humana, Clifton, NJ).

Koop, C. E. (1991) Personal communication.

Kopelman L., Irons, T., and Kopelman, A. (1988) Neonatologists judge the "Baby Doe" regulations. *N. Engl. J. Med.* **318,** 677–683.

Lorber, J. (1972) Spina bifida cystica — results of treatment of 270 consecutive cases with criteria for selection for the future. *Arch. Dis. Child.* **47,** 854–873.

Murray, T. and Caplan, A., eds. (1985) *Which Babies Shall Live?* (Humana, Clifton, NJ).

Nimz, M. (1989) How protected are infants with disabilities?, in *The Triumph of Hope* (Andrusko, D., ed.), NRL, Washington, pp. 145–157.

Powell, T. and Hecimovic, A. (1985) Baby Doe and the search for a quality of life. *Exceptional Children* **6,** 15–23.

Rhoden, N. (1986) Treating Baby Doe: The ethics of uncertainty. *Hastings Cent. Rep.* **16,** 34–42.

Rhoden, N. and Arras, J. (1985) Withholding treatment from Baby Doe: From discrimination to child abuse. *Milbank Memorial Fund Q.* **63,** 18–51.

Weir, R. (1984) Selective nontreatment of handicapped newborns (Oxford, New York).

Weir, R. and Bale, J. (1989) Selective nontreatment of neurologically impaired neonates. *Neurol. Clin.* **7(4),** 807–820.

Wilkie, J. (1991) Personal communication.

Young, E. and Stevenson, D. (1990) Limiting treatment for extremely premature low-birth weight infants. *AJDC* **144,** 549–552.

Parental Perspectives on Treatment–Nontreatment Decisions Involving Newborns with Spina Bifida

Patricia A. Barber,
Janet G. Marquis,
and H. Rutherford Turnbull III

There can be little doubt that the intense debates initiated and fueled by the cases of Baby Doe and Baby Jane Doe have called out for data concerning the processes by which decisions to treat or not treat newborns with life-threatening birth defects have been made. That was apparent early in the 1980s and remains so even now.

Accordingly, in 1984, a team of researchers at the University of Kansas received funding from the National Institute for Disability and Rehabilitation Research to investigate those very processes. They recognized that, since public and professional discourse focused on the ethical, legal, social, and medical aspects of granting or withholding appropriate medical treatment from such newborns, there was a huge gap in what was known. Extant information was minimal, and there were virtually no data that might inform and shape the funda-

From: *Compelled Compassion* Eds.: Caplan, Blank, and Merrick
©1992 The Humana Press Inc.

mental issues of rights, standards, values, and professional practice. In particular, there were no data about how decision making was actually being experienced by parents of those newborns. Although isolated cases of private decisions were reported by the media and by professionals who were willing to enter the debates about regulation or nonregulation of decision making concerning medical treatment, little was known about how the decisions were in fact made, and few attempts had been made to document the parents' perspectives (Darling, 1979).

Nonetheless, recognizing that the decision-making process and the persons involved in the process were entitled to (if not seeking) guidance, a coalition of professional and consumer-advocacy groups had developed and disseminated a statement of Principles of Treatment of Disabled Infants. (The third author of this chapter was a draftsman of the Principles). In addition, the United States Supreme Court had struck down the federal regulations, promulgated under Section 504 of the Rehabilitation Act (prohibiting discrimination against otherwise qualified handicapped people solely on the basis of their disabilities), on the arguably unsupported ground that parents, not professionals, make the treatment–nontreatment decision, and thus, there is no factual basis justifying federal regulation of the decision-making process in medical facilities that receive federal financial assistance (*Bowen v. Am. Hosp. Ass'n.*, 1986). Moreover, Congress had enacted the Child Abuse Amendments of 1984, and the Department of Health and Human Services had promulgated regulations to implement those amendments. These amendments and regulations not only responded to the Court's invalidations of the Section 504 regulations, but also relied on and referred approvingly to the Principles of Treatment. Yet there still remained a knowledge gap in the sense that data about the decision-making process were unavailable.

Development of the Research Project

Emerging from this context was a five-year, exploratory investigation of medical decision making for infants born with spina bifida cystica, aimed at documenting:

1. The process by which treatment and nontreatment decisions were made by parents and professionals;
2. Parent and professional roles related to the decision process; and
3. The major factors that affected the outcome of these decisions.

Data were collected between 1984 and 1989 from five separate cohorts:

1. Two sets of parents;
2. Physicians;
3. Nurses; and
4. Adults with spina bifida.

This chapter draws on the information provided by parents of children born with spina bifida cystica regarding their participation in medical decisions to provide or withhold life-saving surgery for their newborn child. The retrospective accounts of these parents are not intended to provide empirical data describing the actual events that occurred when these children were born, nor was there any attempt on the part of the research team to identify and document solely decision processes that led to the withholding of medical treatment. Indeed, such a focus would have provided an incomplete and skewed representation of the "decision-making process" that, to a greater or lesser degree, was experienced by most parents in this study, regardless of their treatment decision.

The following discussion provides a partial account of our many hours of conversations with these parents. It is intended as an introduction to, rather than a synoptic account of, the parental view of medical decision making—their experiences and their beliefs about the nature of the process.

Study Design

Spina Bifida

Spina bifida cystica, or myelomeningocele, is a birth defect characterized by a malformation of the spinal cord. Major disabilities associated with this condition include partial or total paralysis, bowel and bladder dysfunction, and hydrocephalus, with 10–30% of these children experiencing some degree of mental retardation. The initial life-saving surgery consists of closing the "open spine" created by this condition, usually within the first few hours or days after birth, in order to prevent the onset of meningitis, an infection that can further compromise the child's health or lead to death.

The case of spina bifida embodies all of the major issues surrounding medical treatment decisions for newborns with disabilities, in that aggressive medical treatment at the onset will save the lives of most of these infants, but the outcome in terms of lifelong disabilities can pose a moral dilemma for those who must consent to or withhold the initial life-saving procedure (McClone, 1986).

Population

A two-part data collection procedure, consisting of verbal interviews and written questionnaires, was developed for the parent study. Interviews were conducted with a total of 141 subjects (64 male; 77 female) who were identified through hospital-based spina bifida clinics and spina bifida associations located in the Midwest. Included in this sample were 63 dyads (62 married couples who were parents of a child with spina bifida cystica; one dyad consisting of the biological mother and maternal grandmother) and 15 individual biological parents. A total of 78 children are represented by this cohort.

Out of 141 subjects participating in the study, only 127 (90%) completed the written questionnaire. Therefore, the quantitative data presented in the tables and throughout the study are based on the survey responses collected from this smaller cohort. These 127 parents represent 73 children, or 94% of the children referred to in the total study (*see* Table 1).

The majority of children (68%) were 12 years of age or younger at the time of data collection, with 30 children

Table 1
Description of the Sample

	n	%
Parents completing questionnaire (*n* = 127)		
Gender		
Male	57	45.0
Female	70	55.0
Age		
Range 24–68 yr		
Mean 38 yr		
Median 36 yr		
Education		
<12 yr	14	11.0
Completed high school	40	31.5
College — undergraduate work	49	38.6
College — graduate work	24	18.9
Income		
<$10,000	9	7.1
$10,000–14,999	19	15.0
$15,000–24,999	18	14.2
$25,000–34,999	33	26.0
$35,000 and over	45	35.4
Missing data	3	2.4
Race/ethnicity		
Caucasian	116	91.3
Other	11	8.7
Children with spina bifida (*n* = 73)		
Age		
<1 yr	3	4.1
1–5 yr	27	37.0
6–12 yr	23	31.5
13–18 yr	15	20.5
19 and older	4	5.5
Deceased	1	1.4
Range <1 yr–24		

(38.5%) falling in the birth-to-five age group. This latter group is significant in that their births occurred approximately between the years 1982–1987, a point in time marked by the birth of Baby Doe in 1982, and the ensuing debates and decisions regarding national policy and legal intervention.

Methods

The interview process began with an initial open-ended question that allowed parents to describe the events leading up to their decision concerning life-saving surgery. The second section of the interview consisted of a series of open-ended questions designed to gain specific and consistent information from all subjects regarding the nature of, and any major factors that may have affected, the decision-making process.

In most cases, the written questionnaire was administered at the completion of the verbal interview. Because husbands and wives generally had different experiences following the birth of their child as well as varied opinions about the decision process, each parent was asked to complete a separate questionnaire form.

Limitations of the Sample

At the onset, it is important to note that: (1) life-saving back surgery was authorized for all but one child described in this study, and (2) the majority of parents were raising their child with spina bifida within the home. However, the major research question does not focus on the outcome of deliberation—the final decision. Rather, the questions attempted to document the nature and content of the process that preceded the decision to provide or withhold surgery. Indeed, many of the parents were presented with a nontreatment option, and not all were initially protreatment.

Data Presentation

For reporting purposes, the five-point Likert scales used to measure responses on the questionnaire have been reduced to three-point scales, with the two negative responses combined and the two positive responses combined. The middle (neutral or average) response was left unchanged.

The Parental Experience

The study was designed to document the events that occurred after the birth of a child with spina bifida, as well as the parents' interpretations and evaluations of these early experiences. It was the subjective voice that we sought through the interview process—a voice that had shades of gray that could not be captured by numerical data. This discussion weaves together the separate, but related, story lines created by both sets of data as a means of recreating selected facets of the decision-making process.

Parental Choice
Regarding Life-Saving Surgery

The questionnaire data (Table 2) indicate that a "clear nontreatment option" was presented by the attending physician to at least one parent of 29 children (40%) represented in this study. In lieu of a clear nontreatment option, the parents of an additional 14 children (19%) noted that, through communication with the physician, they were cognizant of the fact that a decision to withhold medical treatment was a possibility. Nontreatment options were not presented for the remaining 30 children in the study. Therefore, one or more parents of over half (59%) of these children were informed of an alternative to life-saving medical treatment.

Further analysis of the data provides a comparison of the percentage of nontreatment options presented for children in two subgroups: (1) those born during the Baby Doe "era" (approximately 1982–1987), and (2) those born prior to 1982. Parents report that physicians presented nontreatment options for 60% of the children in the Baby Doe cohort and 59% of the older children. Therefore, no difference can be found between the percentage of options presented for Baby Doe and preBaby Doe children in this sample.

The selection of the time frame between 1982–1987 is wholly arbitrary on the researchers' part. One could easily

Table 2
Nontreatment Options—Children with Spina Bifida

Nontreatment option presented	Subgroup birth to 5 yr *n%* of subgroup		Subgroup 6–24 yr *n%* of subgroup		Total sample *n%* of total	
Clear	12	40.0%	17	39.5%	29	39.7%
Assumed	6	20.0%	8	18.6%	14	19.2%
None	12	40.0%	18	41.9%	30	41.1%
Total	30	100.0%	43	100.0%	73	100.0%

argue that events preceding the Baby Doe case might have influenced medical practice. With equal fervor, one could defend a dividing line that reflects the actual promulgation of the Child Abuse Amendments in 1984. The major caveat, of course, is that these data do not provide any information about causality. Thus, no current explanation exists as to why the percentage of nontreatment options presented to these parents has remained stable over time. Yet it appears that neither the legal or political climate of the 1980s nor the improvements in medical technology and resources for children with spina bifida over the years have affected medical recommendations regarding life-saving treatment.

Defining the Option
from the Parents' Perspective

Potentially, life-saving back surgery could have been withheld from 43 of the 73 children (59%) represented in this study, whereas a final nontreatment decision was made for only one child. In most cases, parents chose between surgery or "letting the child die," but there were variations other than allowing or withholding life-saving surgery. For example, the recommended treatment for some parents who elected to save the child's life was one of waiting days, weeks, and in one case years before back closure was performed. Yet regardless of the nature of the information presented, the parents' role in the decision process rested on their interpretations of these options and the degree to which they valued the choices they were given.

Parents defined the presentation of these options in a variety of ways. Where they believed that there was a moral obligation to save the child's life, some parents dismissed the presentation of a nontreatment option as little more than the customary medical procedure of requesting permission for surgery:

> [Mother]: I guess he did give us an option. I guess, really, he did, even though we didn't look at it as an option. He's just saying to us, "If you want to give him the best possible chance, then you operate." So, I guess we could have said, "But, what if he's handicapped?" I guess we could have said that, but it never occurred to me—not really.

Others seriously questioned the moral approach of the physician who either suggested or recommended an alternative to life-saving surgery:

> [Mother]: He says, "These are our two options. We could either do the surgery and close up the back, . . . or we could just leave him like this and chances are he would get..." Of course you're going to do the surgery. What do you mean you're not going to treat him? There was just never a doubt in my mind; there was never a hesitant moment. I just couldn't believe the doctor would say that. I was so stunned that he would offer me a choice.

> [Mother]: But no doctor should be allowed to come in and say, "We can do this and that, but I'm not doing it, or you shouldn't do it."

> [Father]: Or, "If it was my kid, I'd let her die." They should never have come in and said that.

An additional group of parents reported little hesitancy in deciding to authorize life-saving treatment, yet their discussions indicate a greater recognition of the moral choice presented to them:

> [Father]: The main thing he [doctor] talked about was, if they didn't operate on her pretty quick and sew up her back, they might as well knock her in the head and throw her in a ditch, because she wasn't going to make it... So, there isn't very much of a decision to make. You either want to keep the child and hope that everything works out, or you don't—you want to give up and start over [with] another one. And a person could have a lot of second thoughts if you just, "Well, I want out of this. I want one that's in good shape."

> [Father]: He said, "The choice is yours. We can do the best we can and try, or we can let nature take its course." The picture he painted was pretty bleak and it almost sounded like an attractive alternative, just to let things go.

Most of these parents rejected both the alternative to withhold treatment and, for some, the negatively biased medical and quality-of-life information they were forced to rely on as the basis for their decisions:

> [Father]: We had doctors with very negative viewpoints, but we still went ahead with the idea that he should have that treatment.

> [Mother]: I had one doctor come in and tell me that he'd be nothing but a paralyzed vegetable.

> [Father]: [Doctor said], "If we treat, then quite likely she'll have a very difficult life. It'll be a life of constant care. Somebody would need to be around all the time."

The highly bleak picture painted for some parents was insufficient to overcome their prolife position. These parents did not see the prognosis for the child as dire enough to warrant a course of action other than the provision of life-saving measures.

Only a few participants engaged in a protracted deliberative process related to nontreatment, with one set of parents actually rejecting the protreatment recommendations of the physician:

> [Father]: Well, I don't recall how the decision came up or how the mention of the closure came up.
>
> [Mother]: But, they kept wanting us to okay it, and we said, "No," initially, that we wanted to think about it.
>
> [Father]: I think what I did, what I felt at that point in time was that if we didn't draw, if we didn't really set up a very aggressive boundary, that we wouldn't have any decision space at all. And so what I went to is probably a much more aggressive "No" than I normally would do. But I felt that until I understood where we were, and until we had a feel for what we wanted to do, I wanted to retain as much ability to make a decision as possible. So [wife] and I basically said, "Look, where we're at right now, this is a disaster, and we're not going to authorize the closure," which is what freaked, got this pediatrician really concerned.

What we see at one level of analysis is data indicating that several parents were presented with clear nontreatment options. Yet the interview information leads to a qualification of these results, insofar as many parents did not define "option" in terms of actual choice. That is, the option held little or no value for many parents, because they made decisions based on previously formed rationales that dealt more with their own moral posture than with physician assessments of medical outcomes.

An additional survey question speaks to the importance of being presented with a nontreatment choice. Out of the total 127 survey respondents, which included those who were not asked to consider withholding treatment, one-quarter (24%) of the parents thought that their physician should have presented a nontreatment option, even though they might have rejected this alternative, whereas the majority of parents (56%) did not believe it was important to be given a choice (neutral = 20%).

The Nature of Informed Consent

Whereas most parents who were given choices neither considered nor acted on them, the questionnaire data still suggest that the rate of options presented to these parents did not diminish over time. What does appear to be changing is both the nature of the physicians' informing interviews and expectations for parental participation in the decision process. An examination of the interview data from parents who were presented with clear nontreatment options depicts some of these changes.

Parents of older children were often presented with highly descriptive, albeit negative, information about their child's condition, and relatively blatant recommendations about the child's treatment:

> [Mother]: He mentioned, "Now, the baby wasn't expected to live and the head would be the size of a watermelon, and the baby would never be able to walk or play or go to the bathroom, be retarded, and there were no chances of her ever surviving." If by some miracle she were to survive, we would have to send her away, and she would have to be instituted for the rest of her life, period.

> [Mother]: And he came in and he told me, he says, "Well, the surgery can be done, but I won't do it. If it was my baby, I would just let her die."

> [Mother]: If you do not do that (back closure and shunt), within six months she can die, either from spinal meningitis, or if you don't put the pump in her head, her head will get so large until she won't be able to do anything, just lay there, and her head will rupture and she will die. That's what we were told. So, at least you can give the doctors credit for being up-front.

A different picture emerges for several parents of younger children: one that is both promising and unsettling. On the one hand, it is obvious from some parental accounts that the highly biased informing interview is giving way to a more neutral and balanced presentation of information:

> [Father]: He told me everything that could happen—the gloom—but, it had a good side as well. I don't think, in any way, he tried to sway my decision on what to do...Yeah, there wasn't anything like, "You have to do this, or..." It was all up to us whether we would do it or not.

On the other hand, there is a parallel trend that may have negative connotations for parents—an approach that may leave some parents confused about their options and, thus, their abilities to make informed decisions:

> [Mother]: I was led, nothing was ever said, and it was never really insinuated that, but there was just something kind of hanging in the air. That if we had decided not to have the operation, that it would be taken out of our hands...It was just more or less the impression I felt.

> [Father]: I think it came through that, certainly, he was recommending that everything medically reasonable be done. I got that impression anyway—the feeling. I'm not sure where I got it from.

> [Mother]: I didn't get that impression...But, I got the impression that he thought it was really pointless in her particular case, because she's pretty severe as far as spina bifida children go, to even do anything.

> [Father]: I don't think that anybody ever said that, leave her there and let spinal meningitis set in and, you know, take its course. Nobody ever really said that, that you didn't have to have surgery. You just kind of read between the lines.

Not only did parents report vague and confusing information about their options, but the ramifications of nontreatment also were not clearly articulated to all parents:

> [Mother]: Well, besides that, it was never discussed
> that much...There is not a humane way to let one of
> these children die. I don't know who's out there and
> who's filling anybody full of what, but...we're not talk-
> ing about a humane death.

> [Father]: There's nobody on that side to tell you what
> can happen if you don't do anything.

> [Mother]: And we've worked with so many other
> parents...I know that's what they're wanting.

> [Father]: So he let me know what it was, various cases,
> and there was a choice. You could do something about
> it, or you didn't have to. That's just the way it was.
> They didn't say it in any way that was life-threatening.

The father of a 15-year-old son provided the following account of the changes he has observed in professional
approaches over the years:

> It's a different kind of thing than it was ten years ago.
> Eighteen years ago the doctor come in and said, da,
> da, da...The doctor no longer comes in and goes, da da
> da. He has to have a different approach...They're
> teaching them better ways of communicating with
> patients. But, if you listen to the content, most of it is
> just teaching them never to put anything in black-
> and-white. Nothing is clear-cut. Continue to offer
> options and let somebody else make the decision. By
> the way you slant the options, pretty well you can guar-
> antee the decision or come close to guaranteeing them.
> But that's the new approach to it. It's just, it's subtler
> and it depends on everybody in the decision process
> having the same amount of information available to
> them. If they do, there's nothing wrong with it. When
> you get into trouble is when the parents usually don't

have anywhere near the amount of information that
the doctors are using for information. So, they're making a decision in the dark based on voice inflections
and indications from a doctor. They really can't
understand what the implications of the decision are.

Although it is widely accepted that shared decision making is preferable to professional domination and that physicians should improve their communication with parents,
when taken to extremes, the "new approach" to parental
informed consent may actually erode parental rights and
responsibilities by restricting their access to clear and complete information required for consent.

Building on Experience—
Parental Recommendations

The respondents' transition from relating their personal
experiences to making statements of beliefs and attitudes
about the decision process occurred quite naturally in the
research process. Parents were able to draw on their own
experiences as well as those of other parents with whom they
have communicated over time. Their responses to questions
concerning who should be involved in the decision process,
the nature of parental and child rights, and the role of the
law in decision making are presented below. The data discussed in this section are based on the responses of the total
127 parents completing the questionnaire and interview
information collected from all research participants.

Participants in the Decision Process

Two survey questions requested parents to identify
the individuals, or groups of individuals, whom they would
select as important participants in the decision-making process, as well as those who should be given final authority.
Parents were able to select one or more of the categories listed
in Table 3.

Table 3
Participants in the Decision Process

Question	Participants	n	%
Those who should be involved (check all that apply)	Child's parents	125	98
	Attending physician	104	82
	Enforcers of federal/state law	19	15
	Other medical professionals	44	35
	State-appointed guardian for the child	18	14
	Hospital human rights committee	20	16
	Other	16	13
Those who should have the "final say" (check all that apply)	Child's parents	121	95
	Attending physician	42	33
	Enforcers of federal/state law	17	13
	Other medical professionals	12	9
	State-appointed guardian for the child	12	9
	Hospital human rights committee	6	5
	Other	7	6

Nearly all of the respondents (98%) supported parental involvement in conjunction with participation from the child's attending physician (82%), and 35% of the parents recommended input from other medical professionals. The inclusion of additional medical personnel is customary given the types of anomalies associated with spina bifida, with most parents reporting conversations with, at a minimum, their obstetrician, a pediatrician or neonatologist, and a neurosurgeon prior to their treatment decisions.

Between 14 and 16% of the parents would include enforcers of federal or state laws, an individual who would represent the child's interests, or members of a hospital human rights committee. Other participants suggested by some of the parents were nurses, social workers, psychologists, other family members, parent advocates, parents of other children with spina bifida, and people who will present religious issues for consideration.

Respondents (95%) clearly placed parents in the position of having the "final say" regarding medical treatment for the child, and approximately one-third believed that these decisions should be made jointly between the parents and the child's attending physician. Closer examination of the data indicates that none of the respondents gave the attending physician sole authority without parental input, and there were six participants (5%) who placed the final decision in the hands of individuals other than parents and physicians. In addition, with respect to the "final say," some form of legal involvement was supported by 13% of the respondents. Parents (6%) used the "Other" category for explanatory purposes, rather than to expand on the list of individuals provided.

Parental Right of Privacy

By acknowledging parents as final decision makers, these respondents are affirming their right of privacy in matters related to the family. Although participants might debate the degree to which their decisions were or should be influenced by individuals other than the child's parents, there is some consensus that, ultimately, the responsibility lies with those who must deal directly and over the long-term with the ramifications of the decision:

> [Mother]: I think that that is something that the parents have to make alone, and I think it's probably very rare that they ever make that decision without outside influence. If they could be told all the facts—here's all the pros and here's all the cons—probably the best thing that they could do would be the two of them go off and meditate and get away from everybody: all their parents, all their ministers. Because in the end, in the end those two people are the ones that are going to face the consequences, whatever those are. They have to live with the consequences of whatever decision they make.

Although parents are unequivocally protecting their right of autonomy in medical decisions, it is the ideal or symbolic notion of parental rights with which they concur. In reality, the respondents found it necessary to define parental autonomy in terms of responsibilities and conflicting interests. In doing so, their views began to polarize over the issue of the nature of parental rights.

Disunity occurred within the cohort of parents when they were presented with a question that juxtaposed the rights of the child with those of the parents. Whereas 95% of the participants supported the concept of parents as the final decision makers, there was a different result when they were asked to agree or disagree with the following questionnaire item: "Children with the same degree of severity of spina bifida should receive the same medical treatment, regardless of parental wishes." Almost half (47%) of the parents agreed with the concept of equal treatment for children, 29% believed that parental discretion held greater value, and 23% left the issue of conflicting rights unresolved by taking a neutral stance. It is at this point that parents begin to take a stand somewhere along the continuum that involves conflict between parent and child rights. The following quotes exemplify some of the "prochild," "prochoice," and "neutral" positions taken by the respondents:

> [Father]: You see all this stuff on the news now about Baby Doe or something, that the parents had left in the hospital and refused to sign to do surgery on this and that. There again, it makes you mad to an extent, but it's the parents' decision.

> [Father]: I think it's wrong that anybody can take someone else's life into their own hands...and say, "Well, she won't have a quality life, so we won't treat, and hopefully, and maybe she'll die, and that way she won't have to suffer through life as an invalid." And I think that's wrong. I think they're making a decision that's not their decision to make...There shouldn't be any choice.

> [Mother]: I'd hate to think that somebody that's got a baby like (child) that's going to be able to be somewhat of a normal child...I don't feel that they should say no, hold off surgery, and just let her die...In a way,

I think they should have the right to choose, but in a way, I don't think they should have a right to withhold treatment from somebody who could be fairly normal.

Whereas some parents continued to defend parental decision making even when, as one father noted, "I think there is a wrong decision," the data suggest that almost half of the participants shifted their support from parental rights to one of defending the rights of the child. Yet this was not always the case. Most of these participants actually maintained their support for parental discretion throughout their discussions, but they did so with the inclusion of an important caveat that reflected their beliefs about the nature of parental rights in decision making:

[Father]: It is the parents' decision, but if it's a life-threatening decision to the child, the parent does not know in the future what the child's going to do, or what he's going to be able to do, and then I do not think it's the parents' decision to terminate a child after he's been born, just because they think that he may not be normal.

[Father]: The parents really should have the right to decide what happens to that kid, as long as that kid is not being abused or mistreated, in my opinion...If refusing him treatment was going to endanger his life, that you could actually see it, then I think they should be made to do it.

These participants define parental autonomy in decision making in terms of the parents' responsibility to the child. A common theme that runs throughout many of their discussions is that parents should have the right of decision if, and only if, they make the morally correct choice—one that reflects the best interests of the child. Embedded within their statements is the belief that, although they themselves can be trusted to protect the interests of the child, they do not trust all other parents to do the same; they are not content to allow others to make the "wrong decision." From their perspective, the parental right to choose is a limited one that should be granted to parents who will place the child's interests above all other considerations. It is possible that, by restricting parental rights in this manner, they are, in fact, validating the treatment choices they made and defending the value of their children's lives.

The interview and questionnaire data indicate that respondents fall into approximately three separate groups based on their responses to the issue of child vs parent rights. There were those who would give parents the "final say" regardless of the outcome, those who were unable to take a consistent stand on the issue, and a final group of parents who defined parental choice in terms related to the best interests of the child.

The Role of Legal Intervention

Parental discussions related to the role of the law in medical decision making provide further clarification of the issue of parental autonomy. It was in relation to the concept of legal intervention that parents began to uncover the final dimensions of parental choice.

Rather than to elicit reactions to the newly promulgated Child Abuse Amendments of 1984, survey and interview questions focused on the general notion of legal intervention. This approach allowed respondents to accept or reject the role of law based on their perceptions of how decisions should be made.

Parents were asked to agree or disagree with the following questionnaire item: "If parents reach a decision regarding treatment or nontreatment that is not in accord with federal and state laws, the parents' wishes should be followed anyway." Approximately 47% of the respondents rejected the law as a final basis of decision in favor of parental choice, 28% did not agree that parental wishes should be followed, and 24% were neutral. Thus, when respondents were asked to choose between parent and child rights, only 29% defended parental authority, but when faced with a choice between parental authority and legal intervention, almost half of the respondents rejected the latter.

Some parents viewed the idea of government intervention as another example of unwarranted intrusion into private family matters:

> [Father]: There's always someone that wants to poke their nose into someone else's affair. But today, I would fight whoever that outsider is that's trying to make my decision. There's always civil liberties sort of thing—want to jump in there and get some headlines. I resent that wholeheartedly. These are decisions that should be made by the parents.

> [Mother]: I don't see that the government has anything to say about it. Where there is two competent parents, there's not a necessity for law.

Others not only rejected a policy that would usurp parental authority, but perceived government intervention, regardless of its form, as inflexible and insensitive to the intricacies, much less the outcomes, of these decisions:

> [Mother]: What did they call her, Baby Jane Doe? We followed that really closely, because that touched home. But, I thought all those people that were fighting those parents did not have to care for that child.

> [Father]: Say they passed that law, and then they run into a parent that absolutely refuses to take care of this baby. Would they be willing, the ones that passed the law, would they be willing to take that baby and put it in their own home and say, "Now, I got the bills and I'm going to take care of the baby?" If they can't honestly do it, then I don't feel that they have any right, saying, yeah, we ought to make that a law.

Still other respondents did not view any type of legal response as the appropriate mechanism for resolving personal, moral issues:

> [Father]: I really don't believe that that sort of thing should be legislated. I don't think it would benefit us...There's a lot of questions in our society and there are things that can't be legislated, and I feel they are anyway. Moral standards that are imposed that I feel that are just not the place of the government, in my point of view.

This view is punctuated by one father's description of the events that ensued after both parents wrestled with the moral dilemma and made the decision to withhold life-saving surgery:

> [Father]: It was difficult enough to decide what we had for someone to say, "You're a bastard for doing this. We're calling in the police because you've made this decision"—just made me go into knots. And at that point in time I became very angry. I don't like people interfering in other people's lives, and this is where I had this sense of government—someone out there—the watchdog had said, "You're not educated enough; you don't have the information enough to make a decision as such." And it left you with a sense that you were uncaring. In some fashion, I got the impression that we were two professionals, and that we couldn't deal with having an imperfect child. That hurt a great deal after making a very difficult decision...But, somewhere in there was lost a sense of what we were trying to do.

It is not surprising that many parents disapproved of state or federal policies that could undermine the very private process of parental decision making, but this was not an overwhelming rejection, with only 47% supporting parental wishes as the final basis for decision making. There were, in fact, several parents who saw the law, or some type of decision-making guidelines, as a useful tool:

[Mother]: I think we've got a lot of government in our lives—sometimes, too much. But without it, things are out of hand...There are people in this world that get themselves in situations that they get into that they don't have enough sense to make the decisions, and they need the law to guide them.

[Father]: I think it should be law that you have to treat. Whether the parents want it or not, that's another point.

[Father]: But the law, I don't know, it's good because it makes a person think—the parents—what would be best.

The most crucial function of legal intervention was at most to prevent parents from making a decision that would be harmful for the child or at least to force parents to take a serious look at the child's concerns. Therefore, parents who defined parental rights in decision making in terms of the child's best interests found a role for the application of the law in preventing "wrong decisions." These respondents were not necessarily in the minority. Although several parents did not endorse any governmental role in decision making, and although some defended parental autonomy at all costs, there was a general feeling among the participants that children with spina bifida, such as their own, deserve the right to opportunities for medical treatment that would enable them to live.

At the very surface of the issue of parental decision making is the fact that these respondents—parents of children born with spina bifida—supported their right to privacy in matters related to family members. Inherent in this right is the parents' responsibility to their child, or what appears to be a majority vote for the inclusion of criteria related to the "best interests of the child." Their stance was generally prolife and protreatment for children born with disabilities.

For the most part, the respondents viewed legal intervention as a threat to parental autonomy, but where they viewed the law as an extension of their own moral views, some parents considered it a useful mechanism for guiding the moral choices of others. This, of course, is the crux of the argument for many parents: Not all parents can be trusted to make decisions that are in the child's best interests. Therefore, the major disagreement between parents was whether "good" children—those whose lives are worth saving—should be protected from the actions of "misguided" parents. Some continued to believe that it was the parent's right to make these decisions, whereas others saw a moral right to protect the child.

In Defense of Moral Choice

The final component of parental decision making—the core beneath these various layers—focuses on the nature of the moral dilemma. Most parents in this sample did not believe that a presentation of a nontreatment option was relevant to their situation, and most parents who were given a recommendation in support of life-saving surgery did not question this course of treatment. That is, the majority of parents believed that the only moral choice was to preserve the life of their child. As such, they did not define their situation as a moral dilemma.

This is not to suggest that parents totally disregarded their options to withhold treatment or that they made decisions without regard for the rightness of their actions. What is evident in the data is that most respondents did not describe a decision process in which they felt that they had to consider and weigh equally defensible alternatives.

What type of situation would create a moral dilemma for many of these parents? If faced with this dilemma, how would parents resolve the issue of parent and child rights? To what degree, if at all, would parents perceive legal or other guidelines to be useful in making these types of moral decisions? Further investigation is needed to address these questions fully, but the data provide some indication as to how these respondents perceived the decision process when faced with these moral dilemmas.

Each parent, whether in vague or specific terms, was able to articulate an infant's physical and/or mental conditions that might lead them to consider the withholding of life-saving measures, but they could do no more than speculate about the actual decisions they would make if their child had been born with the types of anomalies they described. In nearly all cases, they described conditions of greater severity than those experienced by their children. The issue is not one of where they would "draw the line," but whether these situations would require different procedures for decision making.

The comments of one participant provide a transition between what this mother called the "defensive" stance that many parents took when they supported equal treatment against parental wishes and the moral dilemma that many believed they were able to avoid:

> [Mother]: I remember going into the NICU at [hospital] and seeing a sign that said, to the effect, "If treatment is being withheld, you have the right to call this number and report it." I remember the Baby Doe incident quite vividly because that was about the time that [child] was born. And I felt very defensive about it. I thought, "How could people withhold treatment from a child? My opinion's the right one and everybody else is wrong"... and then you see, we go into the hospital and you see everything. You see children with the most debilitating disabilities...had [child] had a severe mental handicap, we might, I might have felt differently about what happened—might have. I don't know.

Preserving the right of parental discretion in decision making becomes more crucial as parents begin to confront these situations, whereas their concern for safeguarding the lives of "good babies" begins to diminish. Where a clear moral choice was no longer obvious, parents shifted their focus from the outcome to the process, and it is here that they took a strong stand regarding their right to be given an option and their right to choose.

[Father]: It's just that, say she would have been a veg-
etable. What doctor, or any other person has the right
to say that husband and wife, you will take care of
this baby because her heart beats and she breathes.

In these situations, parents tended to reject laws or
guidelines that would assert the rights of one individual over
those of another or that would dictate the moral outcome. "I
think about the moral laws," one father noted, and it is precisely this set of standards that they would rely on when con-
fronted with a life of questionable value for the child or the
family:

[Father]: I think there needs to be a law to protect the
kid's right, in which it would probably have to go to a
court case in a lot of them to determine it...But where
it actually comes down to where a machine was keeping a kid alive, where if you unhooked it, he would
die, that, I think, the law has no business being in. I
think it would be the parents' decision.

[Mother]: And like I said, for the law, they have to
protect the child, too, but don't ever make a law where
the people can't say no—that the child has no hope, or
something like this. I don't know how to write it into a
law, but to me, they put too much emphasis on "we
can cure anything," and we can't. And I wouldn't, if it
happens to me again, I still, it depends on the severity—this is for everybody—it depends on how bad your
child is, and only you can make the decision.

Although only a few parents saw some purpose for a law
that would describe the medical conditions warranting a
nontreatment option, they found it difficult to envision how
such a policy could be written to address the nuances of each
situation. One can only speculate on how they would react to
the line-drawing standards set forth in the Child Abuse
Amendments. If they held true to their beliefs about parental decision making, some might reject the very notion of governmental intrusion, others might concur with the legal

standards, and many might argue for increased flexibility and a greater parental role in the decision process.

In general, where parents envisioned a role of legal regulation in decision making, they primarily saw it as a protective mechanism in situations where there was a clear moral choice to preserve life. It might be said that the law was viewed by some as a means of preserving a standard of morality and social conduct, but where there were no undisputed societal standards, the law lost its purpose.

Parental discussions of the "what if?" situations leave us with many assumptions, especially in regard to the role of the law. It is important to note that the idea of legal intervention was an afterthought for most participants—a concept that originated with the researchers and not the parents. However, a few parents dealt with the reality of the law, as others most likely have done since the time of this study.

Although it is not possible to predict how parents would respond in situations other than those they experienced, it is important to give serious thought to how their beliefs and attitudes about the role of the parent fit within current policies and practices related to medical decision making. If parents continue to defend their ultimate right of decision making and the right of privacy, and reject the imposition of external standards in the very situations—the true moral dilemmas—that the law is designed to address, how are these policies affecting the families? We must begin to view the implementation of legal guidelines from the perspective of those who live with not only the consequences of the decision, but also the process.

Addressing Parental Needs

We make no attempt, at this time, to draw parallels to related research or literature. The picture we have presented in this discussion is incomplete, and any theories about the nature of the decision process must remain tentative as we examine the remaining data.

Of greatest relevance to families will be an examination of the types of services, support, and information they require during the deliberative process. The very nature of the process—highly individualized, informal, and often characterized by inconsistencies—can take its toll on parents. The process appears to work best when there is little or no disagreement about the treatment decision. Where parents either challenge professional recommendations or assert themselves into a professionally dominated process, they may find themselves isolated from the support and information they require at the precise time when these are most needed.

The "new approach" to medical decision making requires greater parental participation in the process, but to what degree are parents prepared to assume this role? Standards and guidelines for professional practice are constantly evolving, and professionals can implement and revise their procedures over time. Parents must assume their role without the luxury of prior experience or knowledge, and they require hospital and community-based support designed to meet their informational and emotional needs and, most of all, to facilitate their expanded role in the decision-making process. Although family support policies and programs will need to address the multiple and ongoing needs of all family members, some of the most crucial needs to be addressed are the least complex:

> [Mother]: But I still had this beautiful baby that I was proud of and that I was excited about. I remember [husband] saying, "I wish somebody would just come in and say congratulations."

Concluding Observations

Two major trends emerge from this data. First, physicians continue to present nontreatment options, although the manner in which they present information to parents has changed over time. Nonetheless, the majority of parents regard physicians' presentations of options to be irrelevant in the last analysis to their own decisions; they make decisions that only to a nondispositive degree are influenced by or linked to physicians' statements of options. One might easily conclude that, despite the evidence in this study that many parents regard physicians' presentations of options as basically irrelevant to their substantive decisions and their decision-making process, there seems to be justification for law that requires that physicians present legally sufficient information relative to medical conditions and prognoses, and that other caregivers provide similar information relative to nonmedical support and care for the child and family. The finding of nontreatment options and lack of clarity of information certainly can warrant a full-information rule. The concern has to be with consistent standards of informed consent, not just for the minority of parents who on the basis of these data seem to need it, but also for the increasing number of people who will, at one time or another, have to make decisions about family members who are at the edges of life. The Principles of Treatment and the Child Abuse Amendments of 1984 seem fully justified on the basis of these data, since both call for the presentation of information to family members on which they, together with others, will rely in making their decisions.

Second, parents want to reserve for themselves, and in most cases, for other parents the right to make the decisions about treatment and nontreatment. This privacy and autonomy-related right is not unqualified: The "good parent" should be able to exercise it in connection with the "defective" child, but the "bad" parent should not have so much of it with respect to the "not-so-defective" child (using the "defective" language of the "defective newborn" debate). This result—reservation of a qualified right of privacy and autonomy—has to give pause to the participants in the debates about state regulation of parental decision making. The Supreme Court's decision in the *Parham* case (*Parham v. J. R.,* 1979), involving the role of due process in the decision concerning institutionalization of minors who have emotional or mental disabilities, does not require—but does permit—states to require full-fledged judicial or quasi-judicial hearings on the parental decision for institutionalization. In that respect, the *Parham decision*—much criticized for not requiring more due-process safeguards for the minors—seems consistent with the data revealed here: a limited role of law in regulating parental-right and parental-autonomy decisions involving their children.

The Supreme Court's most recent decision on parental rights, in the *Cruzan* case (*Cruzan v. Director*, 1990), is more problematic for parents if these data represent a dominant view concerning parental decision making and rights. In that case, the Court held that the federal constitution, under the 14th amendment, does not prohibit a state from requiring clear and convincing evidence that the affected individual (the adult child of the parents) had assented or would have assented to the withholding of artificially provided food and water. In this decision, the Court took a position parallel to its *Parham* decision: The states may enact procedural safeguards and may specify the degrees of process due to the parents and their children. Yet *Cruzan* is problematic, because it opens up the possibility that states other than Missouri (whose statute was at issue in *Cruzan*) will impose clear and convincing standards, or similar standards and restrictions, on parental decision making concerning life–death treatment.

It is risky to take the policy arguments too far on the basis of the data presented here, but it is justified to point out that, on the whole, legal regulation of decision making, although not necessarily dispositive of what is decided or how the decision is reached, is not rejected and, indeed, may fit with the stated values of these respondents. Such clarification is important in that some or any regulation (how much is open to debate) should respect the views of those most directly affected—the parents and their families.

Acknowledgment

This research was supported in part by US Department of Education, Grant #G008435060.

References

Bowen v. American Hospital Association, 476 US 610, 90 L. Ed., 2d 584, 106 S. Ct. 2101 (1986).

Child Abuse Amendments of 1984, 42 U.S.C. sec. 5102(2)([B][3]), with regulations at 45 CFR sec. 1341.5 et seq.

Cruzan v. Director, 497 U.S.—, 111 L. Ed. 2d 224, 110 S. Ct.—(1990), affirming *Cruzan v. Director*, 760 S.W. 2d 408 (1989).

Darling, R. B. (1979) *Families Against Society* (Sage Publications, Beverly Hills, CA).

McClone, D. G. (1986) The diagnosis, prognosis, and outcome for the handicapped newborn: A neonatal view. *Issues Law Med.* **2(1)**, 15–24.

Parham v. J. R., 442 US 584, 61 L. Ed. 2d 101, 99 S. Ct. 2493 (1979).

Rehabilitation Act of 1973, Section 504, 29 U.S.C. 794.

Rationing Medicine
in the Neonatal Intensive
Care Unit (NICU)

Robert H. Blank

Major alterations in the United States health-care system will be necessary in the coming decades if we are to avert
a crisis of immense proportions. Many seemingly unrelated
demographic, social, and technological trends in reality constitute a concatenation that promises to accentuate traditional dilemmas in medical policy making. The aging
population, the proliferation of high-cost biomedical technologies designed primarily to prolong life, conventional retroac-
tive reimbursement schemes by third-party payers, and the
realization that health-care costs are outstripping society's
perceived ability to pay all lead to pressures for expanded
public action. At the same time, social institutions appear
both unable and unwilling to make the difficult decisions in
this area traditionally viewed as outside the public arena.

The constraints on economic resources already apparent in the United States are bound to be compounded by the
confluence of the trends noted above. Even with all that is
being done to contain costs, health-care expenditures are
expected to reach $650 billion in 1990 and $1.9 trillion by

From: *Compelled Compassion* Eds.: Caplan, Blank, and Merrick
©1992 The Humana Press Inc.

2000, representing almost 15% of the GNP (Blendon, 1986). Moreover, annual per capita health-care costs in the United States are almost $2000—the comparative figure for Britain is $700. The increased competition for scarce resources within the health-care sector will necessitate resource allocation as well as rationing decisions. In turn, these actions are certain to exacerbate the social, ethical, and legal issues, and intensify activity by affected individuals and groups (Blank, 1988).

Long-term solutions to the health-care crisis require a reevaluation of traditional orientations toward medicine in the United States. Considerable reevaluation of our social priorities is crucial at three levels. First, we must come to some consensus as to how much of society's limited resources we are willing to allocate to health care. As a society, we continue to expend vast sums of money on goods and services that we can largely agree are less worthwhile than health care, and some that actually are costly to our health. What priority do we place on health care as compared to education, national defense, housing, pets, and leisure activities? How much are we willing to take from these competing areas and transfer to medical care? These first-level "macroallocation" decisions entail politics at its rawest since the funding pie is apportioned on the basis of societal priorities.

Second, once a consensus is reached as to how high a priority ought to be put on health-care *vis-à-vis* other spending areas, the finite resources (money, skilled personnel, blood, organs, technologies) available for health care must be distributed among various areas within health care. Assuming that society refuses to make major alterations at the macroallocation level, hard decisions will be necessary to use these limited medical resources in a just and equitable way. This is the allocation level at which the trade-offs must be appraised and choices made that will drastically revise prevailing assumptions of both the health-care community and the public. Here again, conventional "givens" are challenged—givens that appeared to be reasonable when resources were plentiful, but that now in the era of fiscal constraints are obsolete.

Table 1
Trade-offs in Health-Care Allocation

Preventive medicine	vs	Curative medicine
Improved quality of life	vs	Prolongation of life
Young	vs	Old
High-incidence diseases	vs	Rare diseases
Marginally ill	vs	Severely ill
Cost containment	vs	Individual choice

Table 1 presents some of the clearest trade-offs currently debated in the health-policy literature. Whether contrasting a preventive with a curative or rescue approach, care for the young and the old, or an emphasis on quality of life against extension of life, it is unfortunate that the debate often is framed in "either–or" terms. It would be more comfortable to place high priority on all categories or, alternately, to balance them out in an equitable way. Unhappily, the options are narrowing, and resources dedicated to one category do reduce health resources committed to the others. The stakes in how these resources are allocated among the possible categories, therefore, have become quite high, particularly for those elements of society that lack the personal resources essential to protect their own interests. This constrained situation makes the need for establishing meaningful social priorities in health-care allocation even more critical and forces us to make rationing decisions of some type.

Although allocation decisions are arduous, choices become more painful as they are applied to identifiable individuals. Rationing of scarce resources to individuals occurs in one of several contexts. First, a patient might be a member of a specific disease group that has been denied funding under a particular allocation scheme. Under these circumstances, even if it is reasonable to make the macroallocation decisions and withhold scarce resources for that use, justifying it to the individual who will suffer is an onerous task. A decision to no longer fund level-III neonatal intensive care facilities would place all individuals at need in a similar position. Although an ill newborn would not be discriminated against at the individual level, as members of a group they would be denied treatment because society chose to place its priorities elsewhere. Moreover, unless the law places these individuals in a protected class (rejected by the Supreme Court in *Bowen v Am. Hosp. Assoc.*), such policy decisions likely would be upheld under judicial scrutiny.

More problematic is the second application at the individual level. Here, a newborn has a disease or condition where funding is available. However, it is available only in limited quantities, such that not all affected children can be treated. Such rationing always entails a choice between the claims of specific individuals who are competing for resources defined as limited because society has allocated fewer resources than are necessary to treat all affected individuals. Some criteria, therefore, must be established to determine when to use the resources, and when to withhold or withdraw them.

It is important to emphasize that, although allocation decisions are increasingly likely to be made by public-policy makers, the difficult decisions of applying them to individuals should be made largely by parents and physicians. However, although they bear the primary burden of determining which babies will receive aggressive treatment, these decisions will fall within the constraints of social policy. Until now, these limits have not been clearly articulated or specified to the health providers. The traditional physician–patient relationship and third-party reimbursement, combined with the absence of government budgetary sanctions, provide little incentive for considering as a high priority the economic impact on society of an individual medical choice. The climate of cost containment, as opposed to quality of care or equity, which continues to intensify, will alter the incentive structure considerably.

There has been a clear tendency in the United States to avoid making the difficult decisions regarding the distribution of scarce medical resources. Most often, the "solutions" merely shift costs from the individual to the government or from one agency to another. For instance, Aaron and Schwartz (1984) see a pattern toward reliance of policy makers on prospective reimbursement schemes to solve health cost problems. Although giving the appearance of resolving the problem, these shifts only delay the need to make the hard choices at a later date. As Brandon (1982) notes, however, we are running out of easy options and purported panaceas. Despite these interim shifts in the burden of payment, there is a failure to approach the critical issues relating to the need to establish policy priorities and set limits on the use of certain high-cost medical technologies.

Rationing, then, takes place within the context of prior allocation decisions that create scarcity. One of the problems in attempting to formulate an argument for the rationing of health care is the absence of any consistent or succinct definition of what this concept entails. Often, the term is used to describe a process of differentially distributing resources. By emphasizing the selectivity of distribution, this approach leads to the conclusion that persons are treated unequally in any rationing scheme. Other writers have placed emphasis instead on the aspect of constraint on the consumption of medical care applied at the individual level, so that the "patient does not receive all the care that he or his physician believes would be of some benefit to him" (Mulley, 1981). Meanwhile, the Office of Health Economics (1979) questions the use of the term rationing in the health-care context, and suggests that "triage" or "priority selection" be substituted.

Rosenblatt (1981) distinguishes "traditional" rationing, which applies to the poor, with "new" rationing, which applies to all classes. Moreover, he argues that the term has been broadened to include at least two distinct meanings. First, it refers to the distribution of scarce, high-technology treatment among a class of patients who would benefit. This meaning encompasses efforts to "ration" organ transplants, neonatal intensive care, and other technological interventions. The second meaning refers to efforts designed to discourage the use of ordinary health-care resources, such as diagnostic tests, routine surgery, drugs, and hospitalization. These efforts include attempts to change the reimbursement structure and health-care delivery system in order to constrain the overall costs of health care.

Table 2
Forms of Rationing

Form	Criteria used
Physician discretion	Medical benefit to patient
	Medical risk to patient
	Social class or mental capacity
Competitive marketplace	Ability to pay
Insurance marketplace	Ability to pay for insurance
	Group membership
	Employment
Socialized insurance	Entitlement
(i.e., Medicaid)	Means test
Legal	Litigation to gain access and treatment
Personal fundraising	Support of social organization, church, and so on
	Skill in public relations
	Willingness to appeal to public
Implicit rationing	The queue
	Limited manpower and facilities
	Medical benefits to patient with consideration of social costs
Explicit rationing	Triage
	Medical benefit to patient with emphasis on social costs and benefits
Controlled rationing	Government control of medicine
	Equity in access to primary care
	Social benefit over specific patient benefits
	Cost to society

In the broad sense, rationing has always been a part of medical decision making. Table 2 presents a spectrum of ways in which health care can be rationed. Whether imposed by a market system in which price determines who has and has not access; a triage system in which care is distributed on the basis of need defined largely by the medical community, litigation, or corporate benefits managers; or a queue system in which time and the waiting process become the major rationing device, medical resources always have been distributed according to criteria that contain varying degrees of subjectivity. In almost all instances, rationing criteria are founded in a particular value context that results in an inequitable distribution of resources based on social as well as strictly medical considerations.

One disturbing form of rationing often used in neonatal cases is the use of public-relations techniques or other private fundraising activities to secure the funds necessary for expensive medical treatment. Although one hesitates to criticize any technique that reflects care for a human being, rationing by public relations has deep problems. First, it is deceptive. Although such campaigns frequently are sold on the themes of equity and justice within the context of the health-care system, they can result in inequitable and irrational medical decisions. They introduce a strong element of chance or luck to the decision process, because that is often what it takes to strike the public nerve. The public is likely to support enthusiastically the first case, but interest tends to decline as such instances become routine. Moreover, parents who are aggressive, educated, and skilled are more likely to initiate these attempts, thus working against those who lack the temperament and resources necessary to mount a vigorous campaign for funds.

A second problem with public-relations rationing is that medical criteria are easily obscured by the dramatic emotional appeals. Patients who are considerably worse medical risks and less likely to survive, but who are the beneficiaries of such support, gain access to scarce resources, thus depriving more suitable medical candidates of the opportunity. Although one would hope that medical criteria still define the pool of adequate patients for particular procedures, rationing by public-relations appeals is certain to be haphazard and inconsistent. Furthermore, when in conflict with medical criteria, the weight of public pressure is likely to be predominant, resulting in inefficient uses of limited resources. It is not uncommon for medical decisions not to treat to be reversed by public outcry at the denial.

A third problem with rationing by public relations is reflected in the mass media's fascination with the dramatic, the emotional, and the unique. However, these dramatic cases usually involve high-cost curative medicine. They emphasize the technological-fix mentality—the idea that money will solve the problem as well as soothe the egos of those who have contributed to the appeal to save a life. In the process, the media obstructs needed attention on the less dramatic, but equally disheartening cases. Sympathy is easily evoked for the infant who needs a new liver, but not for the corresponding thousands of developmentally retarded children who need special treatment to maximize their potential. Unless one believes there is an unlimited trough of public compassion for giving money to charitable appeals, one is quickly struck with the conclusion that these appeals eventually are in direct competition with each other.

The relevant question today is not if we ought to ration health care, but rather who ought to have prime responsibility for doing it. As illustrated in Table 2, some forms of rationing infer or necessitate government involvement, either direct or indirect, whereas others fail to distinguish between private- and public-sector choices. This distinction, in fact, is very critical to a clarification of how current health-care options differ from those in the past. If the less explicit agents of rationing continue to deny the existence and need of rationing, and their role in it, it is likely that more explicit government-controlled rationing systems, such as the one recently proposed by the Oregon State Legislature, will proliferate.

Policy Issues
in Rationing Medicine for Newborns

Although United States society has moved surprisingly fast toward acceptance of forgoing aggressive treatment at the end of the life cycle (Weir, 1989), it has been considerably less willing to withhold available treatment from newborns who are very ill or premature. To some extent, this reflects a tacit understanding that, although the elderly terminal patient has lived his or her life, the newborn has not yet had the chance to experience life. A second reason why we hesitate not to use all available resources to treat the tiny "defenseless" babies aggressively is their inability to give informed consent. These patients are so utterly, transparently, completely dependent. With no imaginable say of their own, they have been pushed into life with huge burdens already (Menzel, 1989).

Although in theory we talk about substitute consent or a "rational person" standard, no one finds it easy to make a conscious decision to withhold or withdraw treatment, and allow the infant to die. The burden of proof, and often the accompanying guilt, is placed on those persons who would choose to terminate treatment and, thus, end the life.

Within this value context, it becomes even more inflammatory to introduce the dimension of cost when discussing strategies for limiting aggressive, and expensive, treatment for some very ill newborns who have little chance to experience a human life. Many persons argue that cost considerations must not enter these decisions, and any efforts to introduce them are branded as putting a value on the life of the baby and, thus, morally repugnant. As stated in one letter to the editor in the *Chicago Tribune* (June 7, 1990, Sec. 1, p. 26):

> How can anyone put a price on the life of a premature infant? My daughter was 2 lbs., 1 1/2 oz. at birth and is now 3 months and doing great. How dare anyone say it was too costly to save her?

> Any amount of money it costs to save a life is money well spent. Doctors are not gods and medical costs are meaningless compared to saving a preemie's life.

Interestingly, the author of the letter did not state what the final cost was and, more important, who paid it.

Certainly, neonatal care has become the most salient area in which the rapid development of medical technology has produced explosive political issues. With medicine's expanding capacity to save the lives of very ill newborns over the last two decades has come excruciating decisions as to whether to use these capacities or not and who ought to make the decision. Robert J. Haggerty, then vice president of the American Academy of Pediatrics, was correct when he asserted that who decides how to treat newborns with severe disabilities and how those decisions are made "has become the moral issue of our time." It has, unfortunately, also become a political issue of our time. As evidenced by the continuing debate over the Baby Doe regulations and the Child Abuse Amendments, any efforts to ration the use of these technologies will elicit strong, emotional responses from a variety of interests.

Although the policy of requiring aggressive treatment of very ill newborns has been attacked on many grounds, including its adverse impact on parental autonomy, two major issues are critical to the focus of this chapter. The first argument is that our use of aggressive technological intervention to overcome the natural course of events often results in a very low quality of life for the survivors. Ironically, it is the tremendous success of medicine to treat these very ill newborns, itself, that has produced the controversial policy dilemmas. Until very recently, nature largely did the selection as to which infants would survive and which would die. Today, however, increasingly sophisticated neonatal treatment is altering the course of nature. A variety of physical conditions and very low birthweights that in the near past would have been fatal are now treatable. In some cases, the results of aggressive neonatal treatment are good, and the infants survive to lead fulfilling lives. However, in many cases, the quality of life of the surviving infant is so low as to raise doubts about the use of the life-saving intervention. The result, according to the President's Commission (1983), is that medicine's increased ability to forestall death in seriously ill newborns has magnified the already difficult task of physicians and parents who must attempt to assess which infants will benefit from various medical interventions and which will not. Not only does this test the limits of medical certainty in diagnosis and prognosis, it also raises profound ethical issues.

Furthermore, these ethical issues soon become policy issues because, for reasons discussed below, a growing proportion of the bill is paid for by public funds, and because neonatal care is extremely expensive and, in many cases, of marginal long-term benefit. The quality-of-life, sanctity-of-life debate will intensify as we gain the technical capability to "save" even more seriously handicapped individuals.

The second policy issue, which to date has been less commonly articulated, but is at the heart of the problem of neonatal care and overlaps with the first issue, is cost. Simply put, can we afford to continue to expend large amounts of scarce medical resources on infants who at best have a very poor chance of living a meaningful human life? Otherwise stated, is the large potential investment of societal resources in a single person the best use of finite resources, particularly if that person is unlikely to ever experience a productive life? To date, most commentators on treating ill newborns have dismissed the economic dimension as an unwelcome subject. Menzel (1989), reports that, even at the height of the heated conflict over Baby Doe, right-to-life organizations and pediatricians publicly announced their agreement on one thing: Costs and scarcity of resources must not determine such decisions. Although we accept the reality that we are not able to save all infants, as a society, we reject the notion that we let some infants die because it costs too much to keep them alive.

As noted above, however, by rejecting rationing in the Neonatal Intensive Care Unit (NICU), we are deluding ourselves. Although unstated, allocation decisions concerning the establishment and continued funding of NICUs increasingly are being made on strictly economic grounds. The high per patient costs and the low survival/quality-of-life potential of many neonatal candidates, especially those infants under 750 grams, force us to face the distasteful need to include cost considerations just as we must do in all other spending areas. Although the nature of the debate over Baby Doe will change, the aggressive treatment of very ill newborns will escalate as a policy issue as the competition for limited resources heightens. As we come to the realization that not everything technically possible can be done for all patients, pressures will increase to make the hard choice not to treat in some cases. Clear conflicts between the technological imperative and cost containment are evident in neonatal care decisions. Furthermore, aggressive treatment of severely ill newborns pits the basic values of prolongation of life at all costs against quality-of-life concerns and determination as to what course of action is in the best interests of the patient.

Finally, the treatment of very ill newborns contrasts curative or rescue medicine and preventive strategies. Often the need for aggressive life-support systems represents a failure to avert injuries that are preventable. Low-birthweight and very premature babies constitute the major proportion of severely ill newborns, and in most cases, these problems could have been prevented using current knowledge. The Institute of Medicine's Committee to Study the Prevention of Low Birth-Weight in its 1985 report estimated that provision of adequate care to the 3,000,000 women aged 15–39 on public assistance in 1980 could have reduced the rate of premature delivery from 11.5 to 9%. For each dollar spent on prenatal care, $3.38 would be saved. By spending $12.1 million for prevention, $40.91 million would be saved, not to mention the emotional and social costs averted. Another study found that $3.66 was saved in hospital costs for each dollar spent for prenatal care (Nagey, 1989).

Moreover, technological treatment of very low birthweight babies is often rescue in nature, not a cure or even care. Although these procedures may be successful in postponing death and frequently extend the duration of survival of the patients, quality and longevity of that life is frequently marginal. As stated by Ehrenhaft et al. (1989):

Because many very sick newborns who previously would have died are now surviving, an increasing rate of handicap might be expected. Today's technologic advances may be simply postponing death from the neonatal to the post-neonatal period for some proportion of infants.

Often, repeated efforts to maintain even that level of existence are necessary over time at great economic and emotional cost. In addition to the monetary costs, the "costs in pain, grief, guilt, and other intangibles cannot be estimated" (Stahlman, 1989). As stated by Callahan (1990), "Is an average life expectancy of five years, in relatively poor health, at a very high cost, a good or bad outcome?"

Costs of Care for Very Low-Birthweight Infants

Kathy Fackelmann (1988) describes the case of Natasha, who was born prematurely and saved by technology. Natasha has been a resident of Cardinal Glennon Children's Hospital in St. Louis for the four years of her life with no prospect of escaping the life-support system that keeps her alive. The $2,000,000 cost of her care has been shared by Medicaid and the hospital. Although the total number of "million-dollar babies" nationwide is undocumented, the burden they place on public agencies is escalating. As Callahan notes (1990), a $2,000,000 child welfare budget for an entire county consumed by only a few patients requiring extended neonatal care poses nearly impossible dilemmas for the officials who must manage such budgets.

Although these extreme cases remain infrequent, the cumulative costs of treating very low-birthweight (VLBW) babies (variously defined in the literature, but usually less than 1500 grams) is staggering. By 1985, over $2 billion a year was being spent on neonatal intensive care for some 200,000 infants. With the costs of NICU reaching $2400 per day (Gustaitis and Young, 1986), it does not take long to run

up large bills, many of which fall on public agencies to pay.
Moreover, 30–40% of low-birthweight babies require rehos-
pitalization (Shankaran et al., 1988).

The costs of care for the survivors of NICU, however,
mask the full costs because, despite the remarkable strides
in saving VLBW infants over the last decades, low birthweight remains a major determinant of infant mortality.
VLBW babies account for 50% of all neonatal deaths, even
though they represent only 1.15% of all live births in the
United States (Lantos et al., 1988). As compared to normal-
birthweight babies, infants with very low weight at birth have
a relative risk of neonatal death almost 200 times greater
(McCormick, 1985). Therefore, despite our value predisposition against considering costs in caring for very ill newborns,
the high costs per neonatal patient in combination with the
low survival rates, especially of those infants at the lowest
ranges, have heightened policy concerns over the utility of
allocating scarce resources for neonatal care.

In a recent study that compared results of intensive care
for infants with a birthweight of less than 750 grams in 1982–
1985 and 1985–1988, Hack and Fanaroff (1989) found that
mortality and long-term morbidity were not improved.
Despite more aggressive treatment, including more caesarean
sections and active resuscitations, survival rates of babies
below 25 weeks or birthweights under 750 grams did not
improve. The only significant difference in the two periods
was an increase in the mean age at death from 73 to 880
hours, raising a serious question as to whether routine
intensive care is justified for the very immature infant. Table
3 shows that the probability of survival remains very poor if
the length of gestation is less than 24 weeks.

Although Wood et al. (1989) found a significant decrease
in early morbidity of low-birthweight infants compared to
those from similar populations before 1986, chronic lung disease was present at 30 days in all infants born after 24 weeks'
gestation, decreasing to 13% at 29 weeks. Moreover, rates of
severe intraventricular hemorrhage ranged from 100% at 24
weeks to 7% at 29 weeks. Ferrara and associates (1989) found

Table 3
Survival According to Gestational Age, 1982–1988[a]

Gestational weeks	Total births	Survival, no.	Survival, %
20	10	0	0
21	25	0	0
22	29	1	3
23	37	3	8
24	51	8	16
25	80	42	52
26	82	52	63
27	82	59	72

[a]Source: Hack and Fanaroff (1989).

that 23% of VLBW survivors were impaired, a figure that remained relatively constant from 1981 to 1987. However, they found this datum encouraging because of the capacity to save babies at gestational ages and birthweights previously considered nonviable. Resnick et al. (1990) note that, although NICU survivors have significant declines in mental development scores, the declines are caused by sociodemographic factors, not low birthweight *per se.*

In their economic analysis of regionalized neonatal care at the Women's and Infant's Hospital of Rhode Island, Walker et al. (1985) concluded that the benefits for delivering care to infants weighing less than 1500 grams surpasses the costs. However, when examining two weight categories (501–1000 grams and 1001–1500 grams), they found significant differences. Although the economic gain of caring for the larger infants showed "a striking positive advantage," the care of infants under 1000 grams "showed a negative economic effect" (Walker et al., 1985). Although the total benefits/costs for the 1001–1500 gram group was $3,705,988, the benefits/costs for the 501–1000 gram group was $378,774 (Walker et al., 1985).

In another report of the 274 infants weighing between 500 and 999 grams, Walker et al. (1984) concluded that, from the standpoint of cost–benefit analysis, neonatal care may not be justifiable for infants weighing less than 900 grams

Table 4
Cost Components for Very-Low-Birthweight Infants, 1977-1981[a]

Birthweight, grams	Survivor nonsurvivor	Hospital cost/ survivor	Long-term care cost/survivor	Total cost/ survivor
500–599	0/15	—	—	—
600–699	1/37	167,324	192,270	362,992
700–799	19/60	61,792	53,142	116,221
800–899	19/31	41,797	58,388	101,356
900–999	39/26	31,835	7,788	40,647

Source: Walker et al. (1984).

at birth. Table 4 summarizes the data from this study. The neonatal mortality was 68%, with all but one of the 53 born weighing less than 700 grams dying. The total cost per survivor ranged from $40,647 for the 900–999 gram group to $362,992 for the one surviving infant in the 600–699 gram range ($192,270 was in long-term care estimates). In a Canadian study, Boyle et al. (1983) similarly concluded that priority should be given to treating babies in the higher birthweight ranges. By every measure of economic evaluation, it was economically more favorable to provide intensive care for the relatively heavier infants (weighing 1000–1499 grams at birth) than for those weighing 500–999 grams (Boyle et al., 1983).

They go on to argue (Boyle et al., 1983) that any program that does not meet cost-benefit criteria (which include any intensive care for infants under 1000 grams) represents a net drain on society's resources—the program consumes more resources than it saves or creates. Although Boyle and associates admit that rationing of neonatal intensive care, as defined by "preferential provision to those most likely to benefit or to those from whom society is likely to benefit," raises important ethical issues, they nevertheless recommend that countries with limited resources may wish to give priority to infants weighing over 1000 grams.

Lantos et al. (1988) contend that it is imperative that physicians know the effectiveness of therapies for various subpopulations in the NICU, because unlike autonomous adult patients, infants cannot be asked to accept or reject intervention on the benefits and risks of treatment, or the consequences of nontreatment. One such intervention that seems unwarranted for VLBW infants is cardiopulmonary resuscitation (CPR). One study found that the overall survival rate after CPR until discharge was 14%, with 7% still alive one year after discharge (Willett and Nelson, 1986). Lantos et al. studied 158 VLBW (less than 1500 grams) babies admitted to an NICU, 49 of whom underwent CPR. Not one of the 38 babies who received CPR in the first three days of life survived. Moreover, although four of the 11 infants who received CPR after the first 72 hours survived, three of the four had residual neurologic deficits (Lantos et al., 1988).

According to Lantos et al. (1988), CPR, especially in the first 72 hours of life, should be considered an unproven and "virtually futile therapy." Furthermore, futile interventions that offer no immediate or long-term benefit to the patient may not be considered either medically indicated or ethically obligatory. They conclude that VLBW babies should not be subject to standing orders for CPR during the first few days of life. "Aggressive support could still be given such babies, but the need for CPR would be taken as a sign of impending death, and no additional cardiovascular support would be warranted" (Lantos et al., 1988).

As noted above, in addition to the costs of the NICU, long-term care costs for VLBW infants are frequently high. Shankaran et al. (1988) examined the total medical costs from NICU discharge to three years of age for 60 children at the Children's Hospital of Michigan, 35 of whom had neurologic and/or developmental deficits detected following NICU discharge. They found that the medical costs of raising their NICU infants was $1216 per month as compared to $22 for raising a child at home in their study area (*see* Table 5). Of the 60 children, 34 were hospitalized after discharge for a total of 98 hospitalizations (at least 50% of which were part of the continuum of neonatal problems). Moreover, among the 18 children in the moderately to severely handicapped group, there were 72 hospitalizations, for a total cost of $336,654. Of that amount, 95% was reimbursed by Medicaid and other third-party payers, and 4% was absorbed by the hospital as a bad debt.

The authors conclude that, despite the small numbers of their study, low-birthweight infants continue to be vulnerable to serious illness. "These infants require more care than children of normal birth weight even if they have survived the neonatal period without evidence of problems attributable to perinatal and neonatal events" (Shankaran et al.,

Table 5
Average Medical Costs per Month
of Raising a Child from Birth to Three Years[a]

NonNICU—Midwest region	$ 22.00
NICU graduate with	
No disability	$ 62.60
Mild developmental disability	$ 414.20
Moderate to severe developmental disability	$ 650.90
Developmental disability and hospitalized	$1216.00

[a]Source: Shankaran et al. (1988).

1988). They argue that it is now both feasible and economically advantageous to implement preventive programs to reduce the low-birthweight rate. Reduction of this rate will reduce not only the costs of initial hospitalization, but also rehospitalizations and other long-term therapy. They also emphasize the importance of looking at the long-term prognosis of the neonatal patient, and providing the infant and family with a "continuity of comprehensive medical, emotional, and social support" after discharge from the hospital (Shankaran et al., 1988).

Developing Principles
of Rationing in the NICU

One approach to treating newborns, then, would be to apply the same cost–benefit criteria as we might to similar levels of treatment for adults. However, there are other factors unique to infants that could lead to the conclusion that we should ration medical care even more vigorously for neonatal patients than for older children or adults (Menzel, 1989). As noted earlier, some aggressive treatment of very ill newborns is rescue in nature, in that it prolongs the life, but does nothing to improve it. This, of course, applies to life-saving treatment of many adults as well, although there we seem to be more willing to withhold or withdraw such treatment, especially when the patient is terminal and autonomous

(Weir, 1989). More important, according to Menzel (1989), an infant who dies seems to suffer little of the despair of an older child or adult, and death is no affront to an infant's nonexistent expectations. If, for a newborn, death and dying are thus just not as bad an eventuality or process as they are for others, then apparently, without any discrimination against newborns whatsoever, we could spend less on them per survivor than we do on others.

Although this approach would do nothing to allay the emotional burden of the parents of the newborn left to die without life-saving medical intervention, if the policy limits set were consistently applied, they could be defended as a just means of allocating scarce resources. This policy must be set early in the development of the technology before it becomes widely diffused and, thus, expected as available treatment by parents and physicians.

Daniel Callahan (1990) argues for the inclusion of two principles in making medical decisions to use technological intervention, both of which are highly relevant to the neonatal context. First, the principle of "health symmetry" requires that a "technology should be judged by its likelihood of enhancing a good balance between life extension and saving of life and the quality of the life." It should foster the rounded well-being of the person, not benefit one aspect of the individual well-being at the expense of others. For Callahan (1990):

> A health care system that develops and institutionalizes a life-saving technology which has the common result of leaving people chronically ill or with a poor quality of life ignores the principle of symmetry. The saving of very low-birthweight babies at the cost of a poor long-term outcome is an example . . .

Furthermore, any program that does not provide long-term care as generously as high-technology intervention violates Callahan's principle of symmetry.

A second principle proposed by Callahan that has considerable relevance here is the principle of balance. We should make every effort to ensure that no major social group is, as a group, significantly deprived of benefits available to the population as a whole (Callahan, 1990). For Callahan, it makes considerably more sense to focus on disadvantaged groups rather than individuals.

Prenatal care for poor women is, for instance, a more sensible priority to reduce the number and damage of low-birthweight babies than the expansion of neonatal medicine to cope with babies born under harmful circumstances. Meeting the group needs of poor women would be a justifiable priority even if this meant curtailing services for those individual babies (Callahan, 1990).

The practical problem with the principle of balance in our value system with its emphasis on the individual is that we find it more difficult to allocate scarce resources for preventive, aggregate-oriented purposes than for individual care and treatment. Otherwise stated, our strong predisposition against denying aggressive life-saving treatment, especially to newborns, makes this shift in priorities politically unfeasible. We are emotionally unable and unwilling to sacrifice the identifiable individual for the many nonidentifiable individuals who might benefit in the long run. The logically inconsistent policy of the Reagan Administration to cut maternal and prenatal care programs that are proven means of averting low-birthweight babies, and mandating aggressive treatment of many of the products of the failure to provide adequate prenatal care is understandable only in light of our focus on the individual at the expense of the group. As such, it is a clear violation of the principle of balance.

This call for a preventive strategy to deal with VLBW infants is by no means new or unique. Recent reports of the National Commission to Prevent Infant Mortality and the Institute of Medicine Committee to Study Outreach for Prenatal Care both call for universal access to prenatal care as a high national priority (Zylke, 1989). Similarly, other observers have noted the importance of such an approach:

> In any effort to reduce the costs of care, the prevention of neonatal problems appears to be the most reasonable strategy . . .

The top priority should be universal access to good
prenatal care, a goal which other industrialized
nations have been much more successful in achieving
than the United States (Guillemin and Holmstrom,
1986).

Despite these calls for action, the preferred approach
today in the US continues to be curative, not preventive. At
present, only 10% of the total spent on health care for children in the first year of life goes to well-baby care, preventive care, and medical care for problems not requiring
hospitalization (Gustaitis and Young, 1986). Our individual-
istic value system and our faith in medical technology to fix
the problems have led, simultaneously, to a high priority on
saving individual severely ill newborns at any price and a
low priority on community-wide efforts at prevention, despite
convincing evidence of their effectiveness.

How Much Is a Human Life Worth?

How much is the life of a "Baby Doe" worth? The
answer, unfortunately, depends on the level of resources
society is willing to allocate to save him or her. Contrary to
the presumptions of the Baby Doe regulations and Child
Abuse legislation, quality-of-life distinctions always have been
made and will continue to be made in this allocation process.
The broader question of how much a human life ought to be
worth also depends on the value assigned life or a particular
life by society. Some find it inconsistent that we are willing
to expend hundreds of thousands of dollars to save one life,
yet we largely ignore the worldwide deaths of millions of children a year from diarrhea and other avertable causes.
Callahan, for instance, contends that we should give priority
to those persons who can be saved by existing technologies
over the quest for the saving of still more lives by the creation of new technologies designed to press beyond present
frontiers (Callahan, 1990). The greater reluctance to treat ill
newborns aggressively found in other Western democracies,

such as Britain and Sweden (Rhoden, 1986), also demonstrates that there is no one single solution to these dilemmas. These countries' more severe and rigid cutoff points for
treatment result from different and more well-defined priorities as to how scarce medical resources might best be used.

The impending crisis in the allocation of large sums of
money for the treatment of severely ill newborns is inevitable, because it is in direct competition with other health
priorities, such as AIDS and long-term care of chronically ill
adults. The question then becomes not how much is a particular human life worth, but what priority do we put on that
life versus other lives. As stated earlier, every decision we
make to allocate large amounts of resources (money, equipment, blood, trained personnel) to save a life usually diverts
those resources from other potentially life-saving treatments,
albeit often less dramatic ones. The decision to save the life
of a very ill newborn by aggressive medical intervention raises
critical policy questions concerning long-term commitment
of society to invest in substantial "downstream" resources
needed to maximize the potential of that life. Until now, we
have shown more interest in "saving" the life than in the less
dramatic, but more difficult, task of caring for those persons
we have rescued. Before we rush into aggressive treatment
of all VLBW infants or other severely ill newborns, we must
ask whether we have the willingness and capacity to expend
the essential long-term resources to care for them and maximize their potential, whatever that might be. We must ask
the difficult questions, not only of who should make the decision to rescue and on the basis of what criteria, but also who
pays the enormous costs, both tangible and intangible.

I agree with Engelhardt et al. (1986) that society must
determine at what point undesirable standards of health care
are simply "unfortunate," but not "unfair," in the sense of
constituting a claim on further resources. The fundamental
question here, however, is what constitutes "society." Even
among those observers who agree that limits on the use of
medical resources must be established, there is disagreement
as to how and by whom the decisions ought to be made. Cer-

tainly, the most ideal means of allocating and rationing health-care resources would be through some consensus, or at least reasonably common agreement, as to the most just distribution. The diversity of interests in our society and the respective demands they place on the health-care system, however, make such agreement impossible.

One criticism of a policy not to use aggressive treatment on VLBW infants centers on the significant differences in VLBW rates by race. Infants born to blacks are almost three times as likely to weigh less than 1500 grams (McCormick, 1985). Although black infants account for 16.5% of live births, they represent 34% of VLBW babies and 28% of all infant deaths. Furthermore, because a larger proportion of black babies are likely to be dependent on public funding, any public policy that denies such funding is likely to hit hardest on these babies. Although these figures represent a serious indictment of the health-care allocation in the US and reflect much deeper social inequities, a better solution would be to increase spending to upgrade education and prenatal programs to reduce the rate of VLBW infants among black women than to spend large amounts of resources on often futile efforts to save very premature infants, no matter what race they happen to be.

Although a consensus on how medical resources should be distributed is unlikely, agreement is more possible on the procedures through which society will approach these problems. If we can agree that the procedures are fair, and understand that we are bound by them, specific applications, though difficult, might be perceived as unfortunate, but not unfair. As noted earlier, one of the reasons we as individuals and health providers tend to reject the notion of rationing, or any attempt to withhold treatment, is that there is no guarantee that the resources thus saved will be used fairly or even more efficiently. If I as an individual forgo a liver transplant to save the life of my child, most likely its cost will not be spent on prenatal care, but rather on a transplant for someone else—someone who is perhaps less "deserving." The problem then is partly an understandable lack of confidence in the fairness and consistency of present procedures for allocating treatment in the US.

A central element in any effort to ration medical resources in such a way as to be fair yet efficient is education designed to counter the technological imperative. If the parents of newborn candidates for interventions were fully apprised of the risks and side effects of the proposed treatment and counseled as to the quality of life they could expect, some parents would likely remove their children from the pool of candidates. The danger in this approach, of course, is the possible misuse of this education effort by some persons to discourage use of a procedure by exaggerating its negative aspects.

Despite these potential pitfalls, some means must be established to counter the prevailing tendency of Americans to view medical technologies as able to conquer all disease. Without a countervailing emphasis on the risks and dangers inherent in each proposed medical intervention, it is not surprising that we as a society are conditioned to embrace the technologies. As a result, there are many parents who expect and demand treatment only to regret their decision after the resources have been used. Our failure to assess realistically the limits of medicine and the long-term consequences of high-technology interventions, and to communicate this knowledge to the public produces a situation in which we ask for the intervention first and give it serious thought only after the fact. Public expectations must be revised to take into account the limits on what medical science can accomplish, both for society and for the individual.

One means of achieving this would be to move aggressive treatment of VLBW infants from a clinical to research status. For babies born under a prescribed minimum birthweight (e.g., 700 grams) or gestational age (e.g., 24 weeks), parents would be informed that any medical intervention was experimental in nature. Instead of consenting to medical treatment for their very ill newborn, the parents would consent to allowing their baby to take part in a research protocol. For the most premature infants, this approach would better fit the data described earlier than would placing these aggressive life-saving efforts in the clinical context. It would also require a shift in payment from the health-care budget to the research agenda, where it more properly belongs at this point in time.

The "Baby Doe" problem exemplifies the current dilemmas of society regarding the allocation and rationing of scarce medical resources. It also intensifies the need for all concerned persons to participate in a dialog over what the goals and priorities of society ought to be. The trade-offs inherent in any decision must be clarified if we are to appraise the options available to us realistically. Unfortunately, the choices we face are being increasingly constrained by the lack of adequate resources to use all the technologies that allow us to save the lives of all persons.

The heightened debate over neonatal intensive care raises urgent questions concerning the proper use of scarce medical resources. It also introduces the difficult dilemmas of determining the proper role of the government in health-care decision making, and the rights and responsibilities of individuals as they relate to medical care. Finally, it is clear from this area that the trade-offs necessary to allocate medical goods in a just and equitable manner are often painful ones, because not everyone can have the maximum treatment that is technically possible today.

Although many observers today offer rational alternatives for resolving the health-care crisis, decisions made in the health-care arena are unlikely to be founded fully on rational grounds. Health care is a very personal issue for most persons: What might be logical from an economic standpoint on the aggregate level often is unattractive from an emotional perspective of the individual. As Klien (1981) notes, "a great many users" of health care are "hardly the rational shoppers required to make the competitive private market model work." One of the dangers of even positing the need for any type of rationing system for medical care is that, in applying any criteria to specific cases, the outcome appears harsh and even inhumane. This discussion of the treatment of severely ill newborns elucidates the discrepancy inherent in decisions that are deemed rational in the aggregate, but unfair when applied to specific individuals given our prevailing values. These disparities contribute to making health-care policy issues among the most demanding we have yet to confront.

References

Aaron, J. and Schwartz, W. B. (1984) *The Painful Prescription: Rationing Health Care* (The Brookings Institution, Washington, DC).

Blank, R. H. (1988) *Rationing Medicine* (Columbia University Press, New York).

Blendon, R. J. (1986) Health policy choices for the 1990s. *Issues Sci. Technol.* **2(4),** 65–73.

Boyle, M. H., Torrance, G. W., Sinclair, J. C., and Horwood, S. P. (1983) Economic evaluation of neonatal intensive care of very-low-birth-weight infants. *N. Engl. J. Med.* **308(22),** 1330–1337.

Brandon, W. P. (1982) Health-related tax subsidies: Government handouts for the affluent. *N. Engl. J. Med.* **307(15),** 947–950.

Callahan, D. (1990) *What Kind of Life: The Limits of Medical Progress* (Simon and Schuster, New York), pp. 51, 163–167, 197.

Chicago Tribune (1990), letter to editor, June 7, Sec. 1, p. 26.

Ehrenhaft, P. M., Wagner, J. L., and Herdman, R. (1989) Changing prognosis for very-low-birthweight infants. *Obstet. Gynecol.* **74(3),** 528–535.

Engelhardt, H., Jr., Tristram, H., Jr., and Rie, M. A. (1986) Intensive care units, scarce resources, and conflicting principles of justice. *JAMA* **255(9),** 1159–1164.

Fackelmann, K. (1988) Children who need technology—and parents. *Technol. Rev.* **91(1),**26,27.

Ferrara, T. B., Hoekstra, R. E., Gaziano, E., et al. (1989) Changing outcome of extremely premature infants (<26 weeks' gestation and <750 gm): Survival and follow-up at a tertiary center. *Am. J. Obstet. Gynecol.* **161(9),**1114–1118.

Guillemin, J. H. and Holmstrom, L. L. (1986) *Mixed Blessings: Intensive Care for Newborns* (Oxford University Press, New York), p. 283.

Gustaitis, R. and Young, E. W. D. (1986) *A Time to be Born, a Time to Die* (Addison-Wesley, Reading, MA), p. 213.

Hack, M. and Fanaroff, A. A. (1989) Outcomes of extremely-low-birth-weight infants between 1982 and 1985. *N. Engl. J. Med.* **321(24),** 1642–1647.

Institute of Medicine (1985) *Preventing Low Birth-Weight* (National Academy Press, Washington, DC).

Klein, R. (1981) Economic versus political models in health care policy, in *Issues in Health Care Policy* (McKinlay, J. B., ed.), MIT Press, Cambridge, MA, p. 299.

Lantos, J. D., Miles, S. H., Silverstein, M. D., and Stocking, C. B. (1988) Survival after cardiopulmonary resuscitation in babies of very-low-birth weight: Is CPR futile therapy? *N. Engl. J. Med.* **318(2),** 91–95.

McCormick, M. C. (1985) The contribution of low birth weight to infant mortality and childhood morbidity. *N. Engl. J. Med.* **312(2),** 82–89.

Menzel, P. (1989) *Strong Medicine* (Oxford University Press, New York), pp. 97–99.

Mulley, A. G. (1983) Report of the Presidents Commission on Securing Access to Health Care (Government Printing Office, Washington, DC).

Nagey, D. A. (1989) The content of prenatal care. *Obstet. Gynecol.* **74(3),** 516–527

Office of Health Economics (1979) Scarce resources in health care. *Milbank Memorial Fund Quarterly* **57(2),** 265–287.

President's Commission for the Study of Ethical Problems in Medicine and Biomedical and Behavioral Research (1983) Deciding to Forego Life-Sustaining Treatment (Government Printing Office, Washington, DC), 198.

Resnick, M. B., Stralka, K., Carter, R. L., et al. (1990) Effects of birth weight and sociodemographic variables on mental development of neonatal intensive care unit survivors. *Am. J. Obstet. Gynecol.* **162(2),** 374–378.

Rhoden, N. (1986) Treating Baby Doe: The ethics of uncertainty. *Hastings Cent. Rep.* **16,** 34–42.

Rosenblatt, R. E. (1981) Reationing 'normal' health care. *Texas Law Review* **59(4),** 1401–1420.

Shankaran, S., Cohen, S. N., Linver, M., and Zonia, S. (1988) Medical care costs of high-risk infants after neonatal intensive care: A controlled study. *Pediatrics* **81(3),** 372–378.

Stahlman, M. (1989) Implications of research and high technology for neonatal intensive care. *JAMA* **261(12),** 1791.

Walker, D. J. B., Feldman, A., Vohr, B. R., and Oh, W. (1984) Cost-benefit analysis of neonatal intensive care for infants weighing less than 1000 grams at birth. *Pediatrics* **74(1),** 20–25.

Walker, D. J. B., Vohr, B. R., and Oh, W. (1985) Economic analysis of regionalized neonatal care for very low-birth-weight infants in the state of Rhode Island. *Pediatrics* **76(1),** 69–74.

Weir, R. F. (1989) *Abating Treatment with Critically Ill Patients* (Oxford University Press, New York).

Willett, L. D. and Nelson, R. M., Jr. (1986) Outcome of cardiopulmonary resuscitation in the neonatal intensive care unit. *Crit.Care Med.* **14,** 773–776.

Wood, B., Katz, V., Bose, C., Goolsby, R., and Kraybill, E. (1989) Survival and morbidity of extremely premature infants based on obstetric assessment of gestational age. *Obstet. Gynecol.* **74(6),** 889–894.

Zylke, J. W. (1989) Maternal, child health needs noted by two major national study groups. *JAMA* **261(12),** 1687,1688.

Baby Doe and Me

A Personal Journey

Anthony Shaw

What follows is, for the most part, a recapitulation of a personal philosophical voyage, launched as a result of a single patient encounter more than 10 years before the birth of the famous Baby Doe in 1982. In 1970, as Chief of Pediatric Surgery at the University of Virginia Medical Center, I was called to see a newborn baby with esophageal atresia who also had Down syndrome. As the parents signed the consent form, they asked, "Do we have a choice?" I did not respond on the assumption that no response was expected. The baby did well postoperatively, was placed three weeks later in a state institution for the retarded, the usual practice in those days, and six months later died of complications of institutional neglect. The question the parents asked only subsequently struck me as more than rhetorical and led me to pursue this question in an article that was published by the *New York Times Magazine* in January of 1972 (Shaw, 1972).

Going Public

The article, entitled, "Doctor, Do We Have a Choice?" was, I believe, the first article to introduce to the lay public the dilemmas of decision making in the newborn intensive

From: *Compelled Compassion* Eds.: Caplan, Blank, and Merrick
©1992 The Humana Press Inc.

care unit, dilemmas resulting largely from the recent successes of pediatric surgery, anesthesia, and neonatology. In
this article, I raised the question about whether the well-
accepted doctrine of informed consent, when applied to the
situation of the parent acting as proxy for a newborn infant
with mental retardation and a potentially lethal gastrointes-
tinal obstruction, implied a right *not* to consent to life-saving surgery. Although the article excited a voluminous mail
response from the parents of children with Down syndrome,
most of it asserting that their children led happy and productive lives, this article attracted little public attention outside of the *Times* Letters-to-the-Editor column—much less
attention than was accorded by the media to the well-publicized, even sensationalized, case of an infant with Down syndrome and intestinal obstruction, whose death at age 15 days
at Johns Hopkins Medical Center resulted from the withholding of consent for surgery by the infant's parents with the
acquiescence of the baby's attending physicians.

In the following year, 1973, my article, "Dilemmas of
'Informed Consent' in Children" (Shaw, 1973), which
expanded on the issue of medical decision-making for severely
impaired newborns, was published in the *New England Journal of Medicine* in tandem with an article by Yale neonatologists Raymond Duff and A. G. M. Campbell, entitled "Moral
and Ethical Dilemmas in the Special Care Nursery" (Duff
and Campbell, 1973). In my article, coauthored by my wife,
Iris A. Shaw, LL.B.LL.M., I used a series of case reports from
my practice to highlight the increasingly complex dilemmas faced by physicians and parents as a result of the biological and technological advances in neonatal and pediatric
care, dilemmas that were to become major political issues 10
years later.

The hopes expressed in my *New England Journal of
Medicine* article that an approach to the ". . . questions about
the rights and obligations of physicians, parents, and society
in situations in which parents decide to withhold consent
for treatment of their children," might emerge from rational

public discussion did not materialize. Instead, Duff and Campbell's revelations in their companion piece that, of 299 infants cared for in the Yale New Haven Special Care Nursery over a three-year period, 14% died as a result of nontreatment decisions arrived at by their parents and physicians, led to a spate of headlines in the northeastern press and Senate Committee Hearings in Washington, DC. In testimony before Senator Edward Kennedy's committee, Duff estimated that the number of infants dying in the United States as a result of such nontreatment decisions ran into the thousands. However, even these ethically and politically provocative statistics failed to develop sustained public interest.

In Search of Guidelines

Coincidentally, moral philosophers, notably Joseph Fletcher and Paul Ramsey, had turned their attention to the biomedical dilemmas confronting clinicians dealing with perinatal problems. The field of biomedical ethics as we know it today evolved in great part from their writings in the 1960s and 1970s (Fletcher, 1960; Ramsey, 1970) and subsequently from colloquies among ethicists, philosophers, and clinicians, later expanded to involve professionals in the fields of sociology, theology, psychology, and law, and institutionalized in ethics think tanks, such as the Hastings Center, Kennedy Institute, and increasingly, on university campuses nationwide.

During a three-day invitational conference convened in a Pocono Mountain retreat in 1974 by Chester Swinyard, professor of pediatric rehabilitation medicine at New York University, to consider the ethical issues involved in selection of newborn infants with spina bifida for treatment, an international assemblage of ethicists, medical professionals, theologians, sociologists, and legal philosophers considered, in formal and informal sessions, the entire range of issues—medical, moral, and legal—related to decision making for the "defective newborn." (I find it distressing, 16 years later, to recall that only one individual participating in those deliberations over the fate of newborn babies was without an advanced degree and was the only parent present of a baby with the anomaly under discussion by this distinguished group.) In my invited presentation, "The Ethics of Proxy Consent," I expressed my philosophy about proxy decision making for impaired newborns (adumbrating a consensus statement of the President's Commission for the Study of Ethical Problems in Medicine and Biomedical and Behavioral Research nine years later) that "Decisions for or against treatment are best made by doctors and parents together in an atmosphere of full disclosure of information, of support for the family, of concern for the welfare of the newborn baby" (Shaw, 1978). Physicians, I continued, "may be helped by the backup of a multi-disciplined team, usually available in regional centers concentrating specialized services on the high-risk newborn infant."

On the assumption that quality of life (or potential quality of life) was relevant to decision making for handicapped newborn infants, I suggested some guidelines for those assuming the proxy role of decision makers for infants or small children, among which I included a formula (subsequently modified) (Shaw, 1977) intended as a schematic way of conceptualizing the relationships among certain variables that profoundly influence quality of life. The formula, $QL = NE \times (H + S)$ illustrates how quality of life (QL) is dynamically related not only to an infant's natural endowment (NE), including its mental and physical potential with optimal medical–surgical treatment, but also to "social" factors, such as degree of support by its family at home (H) and by society (S). The formula was helpful not only in illustrating how the quality of life of an infant with limited mental endowment (such as a baby with Down syndrome) may be positively or negatively influenced by nonmedical factors, but also in assuring focus on the infant, a theme I developed further in an article published by the *Hastings Center Report* (Shaw, 1977).

I went on to say that "the legislators should reconcile such procedures with child abuse law to give maximum protection to victims of neglect, but at the same time, to remove the threat of prosecution from parents and doctors who choose not to treat. If such freedom from liability is granted to those carrying the proxy of the defective newborn, the law will have to present some guidelines defining situations where nontreatment or passive euthanasia may be acceptable." Under such legislation, I pointed out,

> specific actions or nonactions on the part of physicians resulting in the threat of death to defective newborns will be subject to review by courts, as they now are. Such reviews over a period of time, along with the legislative framework on which court judgements are based, will define social policy in this complex area, hopefully in a flexible way which will help physicians and parents to make humane decisions.

Although the evolution of current policy and practice with respect to decision making for impaired newborns has largely followed this course, much hoped for flexibility and humaneness have been lost in the same kind of rigid, acrimonious polarization that characterizes the abortion debate.

This phenomenon is well illustrated by a handout issued by the University of Wisconsin at Steven's Point in advance of a debate between myself and J. David Bleich, an orthodox rabbi who is professor of Talmud and philosophy at New York's Yeshiva University, in which we were to debate the ethics of nontreatment decisions for incompetent patients, including newborns. The debate was billed as "Life Versus Death!" Inasmuch as Bleich's religious philosophy gave him first dibs on the prolife position, I concluded that I had been assigned the prodeath side! The problem of achieving a rational modus dividend between those who believe that quality of life counts as a factor in medical decision making for both the competent and the incompetent and those who take a vitalist, prolife, sanctity-of-life position, based usually, but not always, on deeply held religious belief may, as it seems to in the abortion controversy, be impossible.

In 1974, in a dual role as chief of pediatric surgery at the University of Virginia Medical Center and chairman of its multidisciplinary Committee for Child Protection, heavily committed to drafting and lobbying for a comprehensive law mandating the reporting and protection of abused and neglected children in Virginia, I discovered that any assumption that parents will make responsible decisions as "proxies" for their infants born with severe handicaps is clearly challenged by the most basic premise of child abuse reporting laws. During one of my appearances before a Virginia Senate Committee hearing testimony on my latest draft of a state child abuse law, a senator, who was also an internist in private practice, pulled the October 25, 1973 issue of the *New England Journal of Medicine* from his briefcase and asked, "Dr. Shaw, how do you reconcile your willingness to accept parents' decisions to let their defective baby die with your support for a law that would require reporting of parents whose actions are likely to result in a child's death?"

I decided to share this dilemma with my colleagues in the Surgical Section of the American Academy of Pediatrics, who authorized me and two colleagues in 1975 to carry out a survey of its members (subsequently broadened to include a national representation of neonatologists, geneticists, and chairs of departments of pediatrics) to determine whether a consensus existed on criteria for nontreatment of impaired newborns and with whom lies the right and responsibility for making such decisions. Among the survey's consensus: Physicians are *not* obligated to save all newborn life; estimated quality of life should be the major determinant of nontreatment decisions; parents first, and physicians second, should carry the responsibility for such decisions with ethics committees ranked third, and courts and legislators ranked a distant fourth and fifth; and finally, the public should/must be involved in these discussions. Of relevance to the future furor over Baby Doe, 76.8% of pediatric surgeons and 49.5% of pediatricians indicated that they would "acquiesce in parents' decision to refuse consent for intestinal surgery in a newborn with intestinal atresia and Down Syndrome." (If the infant also had congenital heart disease, the figures were 85 and 65.3%.)

Although the manuscripts of previous surgical section surveys, all of which dealt with management of specific disease entities, had traditionally been published in the *Journal of Pediatric Surgery*, the *Journal*'s editor-in-chief, C. Everett Koop, rejected our manuscript. Koop, whose role as surgeon general would several years later be central to the dramas of Babies Doe, does not accept quality of life as a criterion for nontreatment decisions (Schaeffer and Koop, 1979). The manuscript reporting the results of the survey was subsequently published in *Pediatrics* (Shaw et al., 1977), the journal for the American Academy of Pediatrics. In addition to the findings noted above, the survey showed that, although pediatricians and pediatric surgeons agreed that parents should be the primary decision makers for impaired newborns, their right to choose nontreatment was a qualified one, related to the severity of the infant's impairment. The pediatric surgeons and pediatricians who indicated that they would acquiesce in a parent's decision to withhold surgery from an infant with Down syndrome and alimentary obstruction (a finding supported by other contemporaneous surveys) (Todres et al., 1977; Crane, 1979) apparently did so as a result of their perception that the future quality of life of such infants would be poor, a perception possibly arising from a skewed experience with institutionalized mentally retarded children.

The survey reflected overwhelmingly a philosophy that each baby should be approached as a unique individual; however, the respondents also felt that guidelines could be established that would help parents and others make decisions in these difficult cases. Many of us expected that such guidelines would be developed by physicians, and those others intimately involved with infants and their families in the medical setting. However, the results of this survey were subsequently selectively interpreted to justify the Draconian

Baby Doe regulations eight years later (Federal Register, July 1983). The guidelines produced so far have been drafted by lawyers, district attorneys, judges, federal bureaucrats, special-interests groups (including physicians), and, most recently, by the US Congress.

Enter the Courts

If most pediatricians and pediatric surgeons agreed that, in certain egregious cases, parents have the right to refuse treatment for their newborn infant based on the potential quality of life of that baby, many judges have disagreed. For example, in the well-publicized case of Baby Houle, an infant born in Portland, Maine, in 1974, with multiple physical deformities, as well as a life-threatening tracheo-esophageal fistula, Judge David Roberts wrote in his decision requiring repair of the fistula in opposition to the wishes of the infant's parents "the most basic right is the right to life itself" (Maine Medical Center, 1974).

During the 1970s, a number of new and growing societal concerns impacted on the nontreatment dilemma in the newborn intensive care unit. These included a wave of new federal and state child abuse laws, increased activism of the prolife movement in the abortion arena, rejection by the aged of life-prolonging technology in living wills, asserting a "right to die," "handicapped rights" activism, and finally, through the efforts of the Hastings Center, Kennedy Institute, and other ethics think tanks, increased public awareness of and sensitivity to problems previously discussed only among medical professionals.

The decade of the 1970s also produced a series of court decisions that established new guidelines for medical management of classes of incompetent individuals other than newborns. Examples include young persons in irreversible coma on ventilators (*In re Quinlan*, 1976), profoundly retarded adults with lethal disease that could be palliated only with painful and frightening treatment (Saikewicz,

1977), the senile and terminally ill on life-support systems (*In re Dinnerstein*, 1978). These and additional cases in the 1980s (*In re Spring*, 1980; *Bartling vs Glendale Adventists Med. Center*, 1986; and many more) in which court decisions placing a "right to die" as a brake on the technological imperative have utilized concepts, some novel and many highly controversial, such as substituted judgment, judicial review, living wills, quality of life, right to privacy, extraordinary means, and best interests.

The first warning to doctors that they might be exceeding their authority in nontreatment decisions, especially with respect to the withholding of surgery, appeared in a 1975 *Stanford Law Review* article by a law professor, John Robertson, entitled "Involuntary Euthanasia of Defective Newborns: A Legal Analysis" (Robertson, 1975). In this article, Robertson argued that physicians who withhold medical or surgical treatment from impaired infants at the request of their parents are vulnerable to criminal charges ranging from child neglect to homicide. Six years later, Robertson's prediction was fulfilled when the attending pediatrician and the parents of a set of severely malformed conjoined twins in Danville, Illinois, were charged—but not successfully prosecuted—with child abuse and attempted murder stemming from a decision to allow the infants to die from lack of sustenance (Robertson, 1981). Although the Danville twins attracted considerable media attention, the medical issues were complex, and public opinion was divided on the ethics of the medical and legal decisions. Baby Doe would change all that!

Come the Babies Doe

Baby Doe was born in Bloomfield, Indiana, on April 9, 1982, with Down syndrome, esophageal atresia, and tracheoesophageal fistula. The child's parents refused life-saving surgery, largely on the advice of their obstetrician; the circuit court judge and the Indiana Court of Appeals accepted the decision, and the baby died six days later. The decisions leading to the death of Baby Doe were widely criticized. President Reagan directed the attorney general and the secretary of the Department of Health and Human Services (DHHS) to invoke and enforce Section 504 of the Rehabilitation Act of 1973, which prohibited discrimination against handicapped persons. This directive led, less than a year later, to the first set of "Baby Doe" regulations (Federal Register, March, 1983), which included a requirement that all hospitals receiving federal funds post the following warning sign: "Discriminatory failure to feed and care for handicapped infants in this facility is prohibited by federal law." The posters urged anyone with knowledge of a handicapped infant being denied "food or customary medical care" to call a "handicapped infant hotline" with a 24-hour, toll-free number. Anonymity of the caller would be protected. Anonymous calls to the DHHS triggered the inappropriate and disruptive descent by federal Baby Doe squads upon numbers of hospitals.

A few weeks after the Baby Doe regulations took effect, they were struck down on procedural grounds by Judge Gerhard Gesell of the United States District Court for the District of Columbia, who called them "arbitrary and capricious," "hasty and ill considered." However, later that year, the defunct Baby Doe regulations were replaced by a similar set of rules (Federal Register, July, 1983) again based on the civil rights of the handicapped, but this time accompanied by the required procedural formalities that were omitted in rushing through Baby Doe I, including an explanation of "need" for the regulations and an adequate period for public comment, the latter consisting of over 16,000 letters of which the vast majority were recruited by antiabortion and handicapped rights groups (97.5% of the letters "favored" the Baby Doe rules).

On October 11, 1983, "Baby Jane Doe" was born in Port Jefferson, New York, with multiple birth defects, including spina bifida, microcephaly, and hydrocephalus. After consulting with physicians, nurses, religious advisors, and others, the parents decided to forgo surgical closure of the spinal opening in favor of "conservative" medical treatment consisting of nutrition, administration of antibiotics, and the dressing of the baby's exposed spinal sac. The DHHS received an anonymous complaint that Baby Jane Doe was being discriminatorily denied medical treatment on the basis of her handicaps in violation of Section 504 of the Rehabilitation Act.

Prodded by the surgeon general, DHHS requested (but was refused) the hospital records in order to determine the validity of the complaint. DHHS took its case to court. The Federal District Court, the Federal Court of Appeals, and finally, the United States Supreme Court ruled that Section 504 did not apply to treatment decisions involving seriously ill newborns and that the government had failed to make a case that nontreatment decisions were being discriminatorily applied on the basis of newborn handicap. Once again, the federal Baby Doe regulations were ruled invalid (*Bowen v. Am. Hosp. Ass'n.*, 1986).

However, the Reagan Administration was not to be denied on what it believed was a Civil Rights issue, and required the United States Commission on Civil Rights to hear testimony in 1985 and 1986 from such interested parties as neonatologists, geneticists, parents of handicapped children, representatives of handicapped rights organizations, and one pediatric surgeon (me). The Commission's report (US Commission, 1989), found in its heavily biased and poorly documented analysis of largely inaccurate and obsolete data what it called "discriminatory denial of treatment" and recommended a chain of oversight mechanisms from local hospital to federal government, which could, if enacted into law, make physicians look back with nostalgia on the Baby Doe regulations of 1983 and 1984.

The report specifically eschews quality of life as a criterion for medical decision making. On page one of the Commission Report, the Executive Summary notes that "in addition to discussing Baby Doe incidents" the report examines "the role of quality-of-life assessments in decisions to withhold medically indicated treatment from disabled infants, exemplified most notably by the quality of life formula,

QL = NE × (H + S), which influenced a team of physicians at Oklahoma Children's Memorial Hospital in denying treatment to 24 out of 69 babies born with disabilities."

The unfortunate use of the word "influenced" in their article, in which criteria for withholding "rigorous treatment" from infants with meningomyelocele were described (Gross et al., 1983) created a false impression that the Oklahoma medical team actually used the formula to determine which infants should live and which should die. Columnist Nat Hentoff reached a similar erroneous conclusion, writing earlier in the *Village Voice* (1984), "a life or death formula . . . used in Oklahoma to speed certain infants to their death." In fact, Gross and his team applied selection criteria based on medical and nonmedical prognostic indicators promulgated by similar groups at other major medical centers (Shurtleff, 1974; McLaughlin et al., 1985). Gross et al. were singled out by the Civil Rights Commission and by Hentoff for clearly stating that their decisions for or against "vigorous treatment" were based largely on quality-of-life considerations with deferral to the informed wishes of the parents of the most egregiously impaired children, i.e., those with severe hydrocephalus, extensive paralysis, or additional severe and unremediable anomalies. Although Gross and his colleagues wrote that they there were "influenced" by my formula, it is clear from a critical reading of their article and from background information subsequently supplied by the Oklahoma Department of Human Services (US Commission on Civil Rights, 1989) that the formula was, in fact, used as I had intended it—"for illustrative purposes" and was never applied (in Hentoff's words) as a "life-and-death formula."

Because of my continuing belief that quality-of-life considerations, as imperfect and inexact as they may be, are not only relevant, but essential to humane medical decision making and because I was concerned that the usefulness of my formula as a didactic tool was being misperceived as an algorithm threatening newborn life, I followed up my 1977 article in the *Hasting Center Report* with "QL Revisited" (Shaw, 1988) which concludes as follows:

> I do not propose this formula as a method of calculating the numerical value of a human life. Nor do I propose it as a guide to a definition of humanhood or personhood, or as a way of assigning points to decide whether lifesaving efforts should be made or discontinued in a particular case. . . . the QL formula identifies in broad terms those factors which affect quality of life.
>
> The QL equation has not relieved nor will it relieve physicians or surrogates for incompetent patients, including impaired newborns, of responsibility for the decisions they make. Ethical medical decision-making requires reflection on the concept of the quality of life, which in turn requires consideration of its constituent factors. The QL equation helps clarify that process in simple terms.

Congress and Baby Doe

Meanwhile, as the Baby Doe regulations took a beating in the courts, Congress was not ignoring this issue. After stormy sessions in committee and on the floor, a heavily politicized compromise passed into law on October 9, 1984, as an amendment to the 1974 Child Abuse Prevention and Treatment Act (Public Law 98-457), based largely on a policy statement signed by the president of the American Academy of Pediatrics jointly with presidents of major disability groups, including the National Down Syndrome Congress, Association for Retarded Citizens, and Spina Bifida Association of America (American Academy of Pediatrics [AAP] Joint Policy, 1984).

The law, which went into effect in October 1985, mandates that, in order to qualify for certain state grants under the Child Abuse and Neglect Prevention and Treatment Act, states must have programs and procedures in place within their Child Protective Service Systems for the purpose of responding to reports of medical neglect, including instances of "withholding of medically indicated treatment" from disabled infants with life-threatening conditions.

> Withholding of medically indicated treatment is defined by the Federal guidelines as: "Failure to respond to an infant's life threatening conditions by providing treatment which [including appropriate nutrition, hydration and medication], in the treating physicians reasonable medical judgement, will be most likely effective in ameliorating or correcting all such conditions. Exceptions to the requirement to provide treatment [but not the requirement to provide appropriate nutrition, hydration, and medication] may be made only in cases in which: 1) the infant is chronically and irreversibly comatose; or 2) the provision of such treatment would merely prolong dying or not be effective in ameliorating or correcting all of the infant's life threatening conditions, or otherwise be futile in terms of survival of the infant; or 3) the provision of such treatment would be virtually futile in terms of survival of the infant and the treatment itself under such circumstances would be inhumane.

The 1984 Baby Doe amendment to the 1974 federal Child Abuse Prevention and Treatment Act places responsibility for reporting and intervening in cases of "medical neglect" involving disabled infants within the Child Protective Services System of each state, and charges the states with developing their own procedures for such cases subject to the broad guidelines in the new legislation. Although not mandating it, the legislation encourages the establishment within health-care facilities, especially those with tertiary level neonatal care units, of Infant Care Review Committees (ICRCs) (Federal Register, 1985).

For the purposes of educating hospital personnel and families of disabled infants with life threatening conditions, recommending institutional policies and guidelines concerning the withholding of medically indicated treatment from such infants, and offering counsel and review in cases involving disabled infants with life threatening conditions.

In the view of the US Civil Rights Commission, as reflected in their 1989 report, these ICRCs were conceived not as "ethics committees," but as hospital internal review mechanisms to assure that the new regulations were being followed in the nation's Neonatal Intensive Care Units (NICUs) and that the unique circumstances surrounding the question of withholding or withdrawing treatment from each impaired newborn were to be irrelevant to the committee's deliberations, which in the commission's view should be limited to guaranteeing physician compliance with the Baby Doe rules. A recent study by the University of Connecticut in conjunction with the AAP shows that most hospitals with NICUs have such ICRCs (Fleming et al., 1990). At the same time, this study indicates that, although these committees are seldom called on for case review, they may, as I have found in my experience as chair of my hospital's ICRC, be helpful in illuminating issues and resolving conflict when the best interests of a severely impaired newborn need definition and focus.

Following the passage of the 1984 Baby Doe amendments to the federal Child Abuse Prevention and Treatment Act, almost all of the states fell into line, incorporating the guidelines for withholding treatment from impaired newborns into state law for protective services. California's reason for holding out (and thus being excluded from about $1,000,000 in federal grants, a modest fraction of the state budget for child protective services) is that its constitution specifically includes "privacy" among the inalienable rights of its citizens. Moreover, California's penal code contains the following language, which would appear to support decisions made by parents and physicians to withhold treatment in appropriate cases: "an informed and appropriate medical decision made by parent or guardian after consultation with a physician or physicians who have examined the minor does not constitute neglect."

Whither Baby Doe?

What of Baby Doe in the 1990s? The Baby Doe of the 1990s is no longer an infant with Down syndrome. For a number of reasons (possibly the result of attitudinal and pragmatic changes stemming from public discussion initiated in the early 1970s), withholding surgery from a baby with Down syndrome with a tracheoesophageal fistula or duodenal atresia is probably unacceptable to most pediatricians, a majority of whom in one recently published study would get a court order for surgery if parents refused consent (Todres et al., 1988) in sharp contrast to the responses reported by Todres et al. in their survey of a similar population of pediatricians 11 years earlier (Todres et al., 1977). Baby Doe now is likely to be a 500–700 gram premature baby on ventilatory support, an infant with short-bowel syndrome resulting from volvulus or necrotizing enterocolitis (Caniano and Kanoti, 1988), an infant with multiple complex physical anomalies coupled with severe brain damage, or an infant in a persistent vegetative state requiring tube feeding and gastric fundoplication to prevent death from aspiration pneumonia. Among the institutionalized mentally retarded pediatric population, the number who are technologically dependent continues to increase according to recent studies (Crain et al., 1990).

There are a number of factors that will influence the efforts made on behalf of the Baby Does of the 1990s.

1. *Cost/Benefit Analysis.* More and more, American society is engaged in an examination of how much it is willing to spend on health services, especially on intensive care for

the elderly and the newborn. The state of Oregon, for example, is planning to prioritize its whole range of health-care services for which public monies are spent, from prenatal care to liver transplants. How will Baby Doe fare in such a policy?

2. *Government Action.* Several recent analyses and surveys in both medical and legal literature are critical of the adverse effect the Baby Doe amendments to the Child Abuse Law may have on decision making in the newborn nursery, which many believe may result in overtreatment and inappropriate experimental treatment of infants who they believe do not meet the strict statutory criteria for withholding or withdrawing life-saving measures (Kopelman et al., 1988; Newman, 1989). According to an analysis by Professor Stephen Newman of New York Law School in a recent *American Journal of Law and Medicine* (Newman, 1989), the "Federal Law fails . . . by not making consideration of benefits and burdens of treatment the lynchpin of analysis for medical decision-making for impaired infants." The extent to which physician behavior is modified by their perception of what Baby Doe laws require of them remains to be seen.

Although the 1983 report of the President's Commission for the study of Ethical Problems in Medicine and Biomedical and Behavioral Research concerning life-sustaining treatment concluded, with certain qualifications, that ". . . parents are the decision-makers for their disabled infants," the 1989 report of the US Civil Rights Commission concluded that legislative action placing impaired infants under the protection of federal law with mechanisms to take decision making out of the hands of parents and physicians is necessary to safeguard newborn life no matter how egregiously impaired. (Of interest in this regard are the experiments at Loma Linda Medical Center, which appear to show that the lives of anencephalic infants may be sustained indefinitely on ventilators, and a recent report from the University of Iowa that as many as 10% of children with Trisomy 18 may survive past the first year of life, and even long-term, with aggressive medical

management [Van Dyke and Allen, 1990]). Families
charged with the continuing management of technologically
dependent infants, some of whom are profoundly retarded,
are often caught between the Scylla of inadequate societal
support in the form of funding and services and the
Charybdis of prosecution under child neglect laws if the
child fails to thrive in the home setting.

3. *The Courts.* What the courts will do in the 1990s is anyone's
   guess. The fact that the federal courts now have a heavy
   representation of conservative judges would suggest that
   courts may support a more rigid prolife stance, and be less
   willing to examine the burdens-and-benefits equation in
   difficult Baby Doe cases.

4. *Child Abuse Legislation.* Finally, there is a great
   inconsistency in the federal Baby Doe regulations as now
   incorporated in one form or another in most state statutes.
   With a handful of exceptions, states have a provision within
   their child abuse law, which, in effect, exempts from a
   charge of child neglect a parent whose child dies or is
   damaged from a curable disease, such as meningitis or
   appendicitis, while undergoing "treatment by spiritual
   means according to the tenets of a recognized religion." The
   contrast between the statutory exclusion of the children of
   Christian Scientists by such "religious exemptions" from
   the protection of most state child abuse laws, coupled with
   the disempowerment of parents with hopelessly impaired
   infants in medical decision making by the same laws is
   one of the unfortunate ironies of the politicization of Baby
   Doe (*see also* Nolan, 1990). Recent criminal convictions of
   parents whose children died from medically curable disease
   under the sole ministrations of Christian Science
   practitioners should impel state legislatures to rescind this
   religious exemption, which some state courts (including the
   Supreme Court of California) have decreed does not shield
   parents from criminal prosecution. It has been almost 20
   years since I wrote in the conclusion of my article "Doctor,
   Do We Have a Choice?" the following:

> Perhaps society should assume more of the deci-
> sion-making responsibility for those who require
> surgery for survival. And if the decision is in favor

of life, society should provide the necessary funds
and facilities to meet the continuing medical and
psychological needs of these unfortunate children.

These days no expense is spared to keep Baby Doe alive
in the nation's NICUs. The challenge for Congress and state
legislatures is to make a wise distribution of resources
between the technological imperatives of the NICUs and the
less dramatic, but no less essential, programs needed to
assure the exBaby Does who carry their handicaps into childhood and beyond a decent quality of life. Many pediatricians
and neonatologists, according to recent surveys, feel that the
Baby Doe regulations compel overtreatment of many of these
damaged infants. If the recommendations of the recent US
Civil Rights Commission Report are written into law, they
would produce intolerable interference with decision making in the nation's NICUs, resulting in increased suffering
at greatly increased costs.

There is a venerable legal maxim that I hope proves to
be true of any new legislation dealing with medical decision
making for impaired infants: *"lex neminem cogit ad vana seu
inutilia peragenda!"*—"the law compels no one to do vain or
useless things!"

References

American Academy of Pediatrics (1984) Joint policy statement. Prin-
ciples of treatment of disabled infants. *Pediatrics* **73,** 559,560.
Bartling v. Glendale Adventists Medical Center, 184 Cal App. 3d 961,
228 Cal Rpts. 360 (Ct. App. 1986).
Bowen v. American Hospital Assn., 476 U.S. 610 (1986).
Caniano, D. A. and Kanoti, G. A. (1988) Newborns with massive
intestinal loss: Difficult choices. *N. Engl. J. Med.* **318,** 703–707.
Child Abuse Amendments of 1984. Pub L. 98-457.
Crain, L. S., Mangravite, D. N., Allport, R., Schour, M., and
Biakanja, K. (1990) Health care needs and services for technol-
ogy—dependent children in developmental centers. *West. J. Med.*
152, 434–438.
Crane, D. (1979) "Decisions to Treat Critically Ill Patients" in*The
Sanctity of Social Life* (Transaction Books, New Brunswick, NJ),
pp. 35–65.

Duff, R. S. and Campbell, A. G. M. (1973) Moral and ethical dilemmas in the special care nursery. *N. Engl. J. Med.* **289**, 890–894.

Federal Register. July 5, 1983 Part III. Department of Health and Human Services. Nondiscrimination on the Basis of Hardships Relating to Health Care for Handicapped Infants; Proposed Rules. **48**, 30846–30852.

Federal Register. April 15, 1985 Part VI. Department of Health and Human Services Part VI. Services and Treatment for Disabled Infants. Infant Care Review Committee Model Guidelines **50**, 14893–14901.

Fleming, G. V., Hudd, S. S., LeBailly, S. A., and Greenstine, R. M. (1990) Infant care review committees. The response to federal guidelines. *Am. J. Dis. Child.* **144**, 778–781.

Fletcher, J. (1960) *Morals and Medicine* (Beacon Press, Boston, MA).

Gross, R., Cox, A., and Tatyrek, R. (1983) Early management and discussion making for the treatment of myelomeningocele. *Pediatrics* **72**, 450–458.

Hentoff, N. (1984) They're putting babies on death row in Oklahoma. *The Village Voice*, 8, May 1, 1984.

In re Dinnerstein, 6 Mass. App. Ct. 466, 380 N.E. 2d 134 (1978).

In re Quinlan, 70 N.J. 10, 355 A. 2d 647 (1976).

In re Spring, 380 Mass. 629, 4065 N.E. 2d 115 (1980).

Kopelman, L. M., Irons, T. G., and Kopelman, A. E. (1988) Neonatologists judge the "Baby Doe" regulations. *N. Engl. J. Med.* **318**, 677–683.

Maine Medical Center and *Martin A. Barron, Jr., M.D. v. Lorraine Marie Houle and Robert B. T. Houle.* State of Maine Superior Court Civil Action Docket No. 74-145 (1974).

McLaughlin, J. F., Shurtleff, D. B., Lamers, J., Stuntz, J. T., Hayden, P., and Kropp, R. J. (1985) Influence of prognosis on decisions regarding treatment of infants with myelodysplasia. *N. Engl. J. Med.* **312**, 1589–1594.

Newman, S. (1989) Baby Doe, Congress and the states: Challenging the Federal Treatment Standards. *American Journal of Law and Medicine* **15**, 1–60.

Nolan, K. (1990) Let's Take Baby Doe to Alaska. *Hastings Cent, Rep.* **20**, 3.

President's Commission for the Study of Ethical Probelms in Medicine and Biomedical and Behavioral Research (1983) *Deciding to Forego Life-Sustaining Treatment: A Report on the Ethical, Medical and Legal Issues in Treatment Decisions.* GPO, Washington, DC, Rehabilitation Act of 1973, 504, 29 USC. 794 (1982).

Ramsey, P. (1970) *The Patient as Person* (Yale University Press, New Haven, CT).

Robertson, J. (1975) Involuntary euthanasia of defective newborns: A legal analysis. *Stanford Law Rev.* **27,** 213–269.

Robertson, J. A. (1981) Dilemma in Danville. *Hastings Cent. Rep.* **11,**5–8.

Superintendant of Belchertown State School v. Saikewicz, Mass. 370 N.E. 2d 417 (1977).

Schaeffer, F. A. and Koop, C. E. (1979) *Whatever Happened to the Human Race?* (Fleming H. Revell Co., Old Tappan, NJ).

Shaw, A. (1972) Doctor, do we have a choice? *New York Times Magazine,* pp. 44–54.

Shaw, A. (1973) Dilemma's of "informed consent" in children. *N. Engl. J. Med.* **289,** 885–890.

Shaw, A. (1977) Defining quality of life: A formula without numbers. *Hastings Cent. Rep.* **7,** 11.

Shaw, A. (1978) The ethics of proxy consent, in *Decision Making and the Defective Newborn* (Swinyard, C. A., ed.), Charles C. Thomas, Springfield, IL, pp. 589–597.

Shaw, A. (1988) QL Revisited. *Hastings Cent. Rep.* **18,** 10.

Shaw, A., Randolph, J. G., and Manard, B. (1977) Ethical issues in pediatric surgery: a national survey of pediatricians and pediatric surgeons. *Pediatrics* **60,** 588–599.

Shurtleff, D. B., Hayden, P. W., Loesser, J. D., and Kronmall, R. A. (1974) Myelodysplasia: Decision for death or disability. *N. Engl. J. Med.* **291,**1005–1011.

Todres, I. D., Guillemin, J., Grodin, M. A., and Batten, D. (1988) Life-saving therapy for newborns: A questionnaire survey in the state of Massachusetts. *Pediatrics* **81,**643–649.

Todres, I. D., Krane, D., and Howell, M. C., et al. (1977) Pediatricians' attitudes affecting decision making in defective newborns. *Pediatrics* **60,** 197–201.

US Commission on Civil Rights (1989) Medical Discrimination Against Children with Disabilities. Washington, DC 20425, p. 366.

Van Dyke, D. C. and Allen, M. (1990) Clinical management considerations in long-term survivors. *Pediatrics* **85,** 753–759.

Baby Doe and Forgoing Life-Sustaining Treatment

Compassion, Discrimination, or Medical Neglect?

A. G. M. Campbell

One outstanding feature of modern medicine is the increasing use of medical technology and its impact on patients—for good and evil (Jennett, 1984). For the most part, this development has brought great benefits that are reflected in improvements to health at all ages, but perhaps most obviously at what Ramsey has called "the edges of life" (Ramsey, 1978). An achievement that has received more attention than most is the ability to rescue a patient from death (as previously defined) and sustain life, at times seemingly almost indefinitely. The benefits of this are obvious. Countless numbers of healthy and productive citizens quite literally owe their lives to medical technology. For some, however, the burdens of survival may outweigh the benefits— those whose lives are severely compromised by disability and handicap, those whose dying is needlessly prolonged, and those in the persistent vegetative state who may have suffered "a fate worse than death" (Feinberg and Ferry, 1984).

From: *Compelled Compassion* Eds.: Caplan, Blank, and Merrick
©1992 The Humana Press Inc.

Neonatal Intensive Care

The intensive treatment of newly born infants is a typical medical success story. It represents triumph of knowledge, skills, and technology when applied to one of the most hazardous times of life. As a result, neonatal mortality rates have fallen to levels hardly thought possible several decades ago and now approach what has been called the "irreducible minimum," if any such thing exists. Morbidity rates have also fallen, and most infants, even those who survive at "the limits of viability," are assured of a healthy future. The increased survival of tiny infants of very low birthweight (VLBW) has been particularly striking (Hack and Fanaroff, 1989). In the 1960s, very few infants survived with birthweights under 1000 grams or at gestation ages of less than 28 weeks. Now, neonatal units report survival rates of 40–50%, including the occasional infant weighing not much more than 500 grams at birth (Harvey et al., 1989).

During the same period, some congenital abnormalities, notably heart defects, that were thought to be untreatable became amenable to corrective surgery, and there have been major improvements in the treatment of other conditions once labeled as "hopeless." With aggressive resuscitation, corrective or palliative surgery, infant respirators, and other life-supporting technologies, the advances in neonatal care epitomize the success of modern medicine.

Like other neonatologists in the 1960s, I promoted and practiced an aggressive approach to resuscitation and intensive treatment as a member of an enthusiastic team of doctors, nurses, and others, all dedicated to the same crusade against death and disease. Such work, although physically and emotionally exhausting, was rewarding in its triumphs and in the joys shared with grateful parents. Success was measured by the saving of infant life. Death was viewed as failure.

However, not all babies grew to be healthy, and not all parents were grateful. Some parents felt that we had become too obsessed with the technical wizardry of high-technology intensive care, and had lost sight of the devastating human and economic costs when survival was compromised by disability and handicap. As the power to restore and sustain life increased, so did the difficult moral questions raised about using this power in all circumstances. We debated, sometimes at great length, the rights and wrongs of using life-sustaining treatment, such as respirators, for some infants. For others, we could not escape the moral and legal uncertainties about decisions to do little or nothing to prolong life, and to "let nature take its course." As we wrote at the time:

> The penetrating questions and challenges, particularly of knowledgeable parents [such as physicians, nurses or lawyers] brought increasing doubts about the wisdom of many of the decisions that seemed to parents to be predicated chiefly on technical considerations. Some thought their child had a right to die since he could not live well or effectively. Others thought that society should pay the costs of care that may be so destructive to the family economy. Often too, the parents or siblings rights to relief from the seemingly pointless, crushing burdens were important considerations. It seemed right to yield to parent wishes in several cases as physicians have done for generations. As a result some treatments were withheld or stopped with the knowledge that earlier death and relief of suffering would result (Duff and Campbell, 1973).

Almost 20 years later, it is appropriate to review these decisions in the light of subsequent events and consider how recent legislative interventions have affected the care of these infants. These dilemmas have not disappeared; in fact, they have increased. Improved prenatal diagnosis with easier access to abortion has reduced the birth incidence of some disabling conditions, like the neural tube defects (a major concern in the 1960s and 1970s), but this decrease has been more than counterbalanced by an increase in the problems posed by infants of VLBW (less than 1000 grams). The immaturity of these infants makes them particularly vul-

nerable to brain-damaging complications, such as hemorrhage, and to the many iatrogenic hazards of invasive intensive care. If pediatricians are to act in their infant patients'
best interests (beneficence and nonmaleficence), decisions to
withhold or withdraw life-saving treatment will continue to
be necessary. There will always be times when an infant's
future *quality of life* may be viewed as just as important as,
if not more important than, the *fact of life* in determining
treatment options.

At this point, perhaps it should be emphasized that "forgoing life-sustaining treatment" or making a "nontreatment
decision" does *not* mean withholding or withdrawing *care*.
More than anything else, making such a decision should
increase the commitment to the infant and family. Infants
are given all the attention and cherishing associated with
high standards of medical and nursing practice—caressing,
warmth, and changing, with appropriate treatment to prevent pain from wounds or technical procedures, and to
relieve distress from hunger or thirst. The withholding or
withdrawal of food and fluids remains controversial, and discussion always raises strong emotions. It is usual for oral
feeding to be continued as accepted and tolerated, but I
believe that, once a decision is taken to allow an infant to die
without life-saving intervention, it is illogical and pointlessly
cruel to prolong dying unnecessarily by starting or continuing intravenous infusions or tube feedings. In each case, of
course, there must be room for discretion according to the
individual circumstances with particular attention paid to
the views and feelings of the parents.

Baby Doe

When several articles describing these treatment dilemmas first appeared in the early 1970s, the initial responses
from medical communities on each side of the Atlantic were
relatively muted. The decisions reported did not surprise doctors and nurses familiar with the tragedies of abnormal birth

or with contemporary practice in neonatal units, but "revelations" that babies were allowed to die certainly upset others. Referring to the Duff/Campbell article. one author wrote:

> An uproar followed the publication . . . not that anyone in medical circles was very shocked that someone had let babies die. Generations of doctors have been doing the same thing on the sly. But here were two doctors openly admitting it (Lyon, 1985).

The great majority of the many letters received by the authors from people in all walks of life were sympathetic and supportive. Much of the hostile correspondence, sadly even from doctors, was based not on careful reading of the original article, but on sensationalized and inaccurate extracts in the popular press. Most who understood the problem acknowledged the need for wider debate, and for considerable latitude and tolerance in coping with individual family tragedies. "I for one am glad that you have faced the decisions as you have and laid the burden open before society which will have to face it too" (Duff and Campbell, 1980). One prominent medical reaction came from C. Everett Koop, the distinguished pediatric surgeon who later became US surgeon general. At a prayer breakfast in Atlantic City, he launched a tirade against us and, in a wildly inaccurate interpretation of what was written, told his startled audience that Duff and Campbell had "starved to death 14% of the population of their newborn intensive care unit over a two year period" (Koop, 1978).

Such polemics and distortions did him no credit, but they gave some early indication of the strength of feeling involved and the kind of tactics later to be used by the so-called prolife movement in campaigning to limit the discretion traditionally given to pediatricians and parents in deciding about the appropriate treatment for severely abnormal infants. Some time later (on April 9, 1982), when an example of apparently inappropriate decision making occurred in Bloomington, Indiana—the original Baby Doe—Dr. Koop used this case and these early articles to indicate why, in his view, legislation was overdue. No evidence of widespread abuse of this discretion by pediatricians or parents was provided.

The initial Baby Doe legislation—"hasty and ill considered;" "arbitrary and capricious" as it was called by one judge—did not fare well in the US courts. The notorious warning notice to neonatal units, the associated directives, the "hot line" access to Baby Doe squads, and the resulting harassment of nursery staff caused considerable distress and probably untold harm if the effects on one unit were any reflection of what happened in others (Shapiro and Rosenberg, 1984). As far as I am aware, this modern witch hunt did not uncover a single case of abuse, despite claims that infanticide was widespread. It must have interfered seriously with the care given to other babies, and undoubtedly caused considerable loss of morale and confidence among the staff and parents. Eventually the Department of Health and Human Services (DHHS) failed in its attempt to bring the care of these babies under Section 504 of the Rehabilitation Act of 1973, which prohibits recipients of federal funding from discriminating on the basis of handicap, a statute that was eventually declared by the US Supreme Court to be inappropriate for the medical treatment of handicapped newborn infants.

However, Congress subsequently amended a child abuse statute that provided that withholding or withdrawing medical treatment in certain circumstances was classified as "medical neglect." To qualify for federal funds, each state currently is required to establish "procedures to ensure that nutrition [including fluid maintenance], medically indicated treatment, general care, and appropriate social services are provided for infants at risk with life threatening congenital impairments" (HR 1904 98th Congress).

US Commission on Civil Rights

In a subsequent development, pediatricians must have been dismayed to see their efforts on behalf of infants and families again labeled as discriminatory. The US Commission on Civil Rights, in responding to Baby Doe, held a series of hearings "to attempt to determine the nature and extent of withholding medical treatment or nourishment from handicapped infants and to examine the appropriate role for the Federal Government."

In a voluminous and one-sided report published in September 1989, the commission indicated its belief that "discriminatory denial of medical treatment, food and fluids is and has been a significant civil rights problem for infants with disabilities. It is also persuaded that the available evidence strongly suggests that the situation has not dramatically changed since the implementation of the Child Abuse Amendments of 1984 on October 1, 1985" (US Commission on Civil Rights, 1989). As suggested in an attached comment, this lack of change may merely indicate that this expensive undertaking by the Commission on Civil Rights was unnecessary, but more seriously, it devalues thoughtful medical decisions taken after considerable debate and with the most honorable of motives—the best interests of the infant—and classifies them as discriminating on the same basis as discrimination on account of race, color, or religion, for which the motives are very different. In my view, Baby Doe and the subsequent legislation have had considerable influence, some good and some bad, and it seems likely that the bad may be made much worse if attention is paid to the recommendations contained in this report.

For example, Recommendation 8 states:

> To create a deterrent to physicians who not only might fail to report a case of withholding to the State CPS agency but also might not submit it to a hospital committee, the P & A [Protection & Advocacy] agency should be given authority to conduct retrospective reviews of the medical records of those with disabilities who die in the State. If instances of illegal withholding of medical treatment are detected, the P & A agency should be able to seek appropriate action by licensing boards or Federal funding sources and, in

extreme cases, to institute suits for injunctive or monetary relief or refer cases for investigation by prosecuting attorneys.

It is worth noting that the chairman, William B. Allen, took the unusual, if not unique, decision to dissent from the final report and expressed the view that "the interests of handicapped newborns have been sacrificed to a political mission." A flavor of this "mission" can be gained from the last paragraph in the statement of Commissioner Robert Destro.

> This report rejects discrimination against the persons with disabilities in the context of medical care decision-making. To have ignored the problem, treated as a matter of medical or parental autonomy, or minimized the seriousness of the attitudes which brought these practices about would have been to take the risk that the public might believe "that [this commission thinks] it is all right to kill kids—[because] that is one way one could interpret [it]."

In a commentary on the recommendations, a physician-ethicist points out:

> there is no evidence of significant undertreating of severely ill handicapped neonates since 1985. If anything there is evidence of overtreatment. Moreover, were there evidence of the failure to provide indicated treatment, the proposed approach would not be the preferred solution to the problem. If the recommendations of the report are aggressively followed the interests of severely ill newborns will be imperiled (Engelhardt, 1989).

Labeling the discretionary treatment of abnormal infants as "infanticide" or as "discrimination against the handicapped" is distressing and offensive to parents and pediatricians who are attempting to act in the best interests of infants and families. It is also grossly unfair to many of those same pediatricians who, throughout their professional lives, have campaigned vigorously for the rights of handicapped children. It is hardly surprising if they feel incensed at the hypocrisy of an administration that labels these actions as discriminatory at the same time as it is cutting funds for services to the disabled (Young, 1984). As noted in 1983:

> Public support for effective voluntary organizations and governmental programs is the inescapable extension of society's deep interest in sustaining life in neonatal units. Furthermore, to the extent that society fails to ensure that seriously ill newborns have the opportunity for an adequate level of continuing care its moral authority to intervene on behalf of a newborn whose life is in jeopardy is compromised (US President's Commission).

In discussing the implications of Baby Doe for current neonatal care, it may be of interest to consider the trends of the past two decades and examine some transatlantic variations.

Intensive Care Since Baby Doe

The United States

From numerous contacts and many visits to the United States, my impression of the current position is consistent with recent reports and analyses (Lantos, 1987; Kopelman et al., 1988; Young and Stevenson, 1990). The final version of the "Baby Doe Rules" allows considerable flexibility in interpretation, but government intervention, the resulting publicity, and the imposition of legal or quasi-legal rules for medical treatment have created a climate of uncertainty, insecurity, and even fear of legal retribution in newborn nurseries. Baby Doe has *not* fundamentally altered the law relating to abnormal infants, nor has it changed the carefully considered practice of many pediatricians, but in some units overrigid interpretation of the "Rules" has led to a feeling of compulsion toward aggressive treatment for all newborn infants irrespective of condition or the future quality of life. This has undoubtedly inhibited some doctors from withdrawing intensive treatment, such as ventilation, even if, to everyone, this seems to be the least detrimental of several unsatisfactory options.

It is acknowledged that determining the prognosis in some cases may be very difficult. In spite of recent major improvements in the techniques of assessing brain damage, some uncertainty will always remain. It is quite possible that an occasional infant who might otherwise have been allowed to die will have been saved by Baby Doe and be healthy, but I suspect that many more infants are being "rescued" only to suffer grievously in their future years.

> The relentless application of intensive care to those neonates whose chances of survival are judged to be "1 in 1000" or "without precedent" is in conflict with the charge of alleviating suffering for the majority of neonates. In such a setting it also becomes difficult to justify the tremendous costs that are incurred. If we fail to introduce some measure of reasoned restraint into our decision-making processes, prolonged suffering and large hospital bills will continue to accompany the uncertainty of our prognosis for a substantial number of very low birth-weight infants (Fischer and Stevenson, 1987).

It is sad, but understandable, if pediatricians simply refrain from doing what they believe to be best for their patient because of concern about their own interests and those of their hospital. Even if the doctor's medical judgment is correct and the criteria for withholding or withdrawing treatment are fully in accord with the Baby Doe rules, there may be someone, from a personal sense of moral outrage or religious zeal, perhaps out to make mischief, or simply from ignorance, who would object sufficiently to "blow the whistle" and alert the State Protection Agency. The resulting investigative process, its effect on the staff and parents, media publicity, and further legal harassment are not pleasant even if all involved are exonerated from wrong doing. From what one hears and sees, one conclusion is inescapable—there are times when the individual whose interests are least protected is the infant himself or herself.

During a visit to a large pediatric intensive care unit in 1987, I was shown an infant with a chromosomal disorder who had been kept alive for over nine months by treatments like ventilation and parenteral feeding at enormous cost, by then approximately $1,000,000. With such expert care, he might survive for many more months. There was no doubt about the diagnosis and no prospect that he would ever lead a normal, or even a reasonably normal, life. His parents, having given up pleading with the doctors to take him off the respirator and let him die, had long since left town and could not be traced. Without exception, the doctors and nurses felt that the time to give up was long overdue. Unfortunately, the Ethics Committee could not agree, so the matter continued to drift. To those most involved, it seemed that his life was being sustained because of pressure from a local prolife group and the inevitable media publicity that would follow death after withdrawal of treatment. I was also told that such publicity might have an adverse effect on the major fundraising campaign currently in progress on behalf of the hospital. The whole scenario was depressing, and particularly upsetting in its effects on the morale of the doctors and nurses who recognized the inhumanity and pointless cruelty of their continued treatment. For parents who witness such prolonged abuse by technology, the effect must be devastating. Such an extreme example may be unusual, but a recent report suggests that overtreatment may be more common than generally thought or acknowledged (Van Dyke and Allen, 1990).

Thus, politically and publicly exploited concern that harm was being done to a few infants through denial of a "right to life" or discrimination against the handicapped has created a situation in which harm is being done to more, perhaps many more, infants through the extensive use of cruel and very expensive treatment (Moskop and Saldanha, 1986). How such a policy can lead to tragedy was apparent from an incident in Chicago in 1989, when a father entered a pediatric unit and, while holding the staff at gun-point, disconnected his infant son from life-support and held him in his arms until he died. "I did it," he said, "because I love my son and my wife." At the very least, this incident must reflect extraordinarily poor communication between the parents and the staff, and is a dreadful indictment of the irresponsible use of intensive life-support and its effect on grieving families.

The United Kingdom

Such tragedies would be unlikely in a European setting. It is doubtful if an infant with, for example, a severe chromosomal disorder, such as Trisomy 13 or Trisomy 18, would have been started on life-supporting treatment. If started, the treatment would have been withdrawn once the diagnosis was confirmed.

The story of Baby Doe received little publicity and has had no significant impact on pediatric practice in the UK. The intervention of the US government and the subsequent legislative actions were followed by British pediatricians with a mixture of sympathy for their American colleagues and abhorrence at the potential effects of such policies on infants and families. There was also some anxiety that similar legislation might prove attractive to the British government, which was also under some pressure from prolife lobbyists as an extension of their sustained campaign against abortion.

Cases identical to Baby Doe have occurred in the UK, and some have come to the attention of the courts. Compared with the US, they have not received such sustained media attention or outpouring of public concern, nor have they stimulated sustained demands for legislative action. It is also true to say that they have not led to any substantial clarification of the law. Indeed, some interpretations have caused further confusion. It is worth recalling that, in 1981, an English pediatrician was tried (and subsequently acquitted) for murder after his patient, an infant with Down syndrome, was sedated and allowed to die (Gunn and Smith, 1985). At the time, a public opinion poll indicated that a large majority were in favor of leaving decisions about abnormal infants largely to the discretion of the parents and their medical advisers. Many thought it was wrong that a pediatrician should face a serious charge in a criminal court for acting out of concern for the infant and family. One reaction to this trial was a proposal for a Limitation of Treatment Bill, which, with appropriate safeguards, would allow the withdrawal of life-sustaining treatment up to 28 days after birth (Brahams and Brahams, 1983). Although such legislation might provide some legal protection for doctors when making difficult decisions to withhold or withdraw treatment, pediatricians generally were reluctant to see any increased state involvement in these matters either to enforce life or sanction death. It was felt that there was considerable merit in doctors and parents continuing to trust each other in seeking the least detrimental treatment options for an individual infant, even if it caused much uncertainty and anguish. The existence of any Limitation of Treatment legislation might erode much of this trust.

In a case very similar to Baby Doe, and also involving an infant with Down syndrome, the parents refused permission for an operation to correct an intestinal obstruction. The first surgeon contacted would not operate without their consent. The Social Services Department petitioned the authority, and the child was made a ward of the court. At the first hearing, the judge upheld the decision of the parents, but this was reversed by the Court of Appeal. The court decided that, because the baby would live "the normal life of a child with Down syndrome" if the operation was performed, surgery should proceed. It is important to emphasize that the judges did not say that, in all cases of handicap, the correct course would be to provide treatment, a point underlined in another important decision in 1989 (Brahams, 1989). In this decision, the Court of Appeal authorized treatment to relieve suffering (for a child with severe hydrocephalus and other abnormalities), but indicated that *it was for the responsible doctors to decide* if they should do anything to prolong the infant's life. This case came to the attention of the Court of Appeal only because the child was already a ward of the court for reasons unconnected with her medical condition.

Mr. Justice Ward concluded that the first and paramount consideration should be the well-being, welfare, and interests of Baby C, but in wording his court order he used an unfortunate phrase, "to treat to die." To allay anxieties that this might be construed as allowing the doctors to bring about death by starvation or an overdose, the case went to the Court of Appeal. In their judgment, the Lords of Appeal indicated that Mr. Justice Ward had "failed to express himself with his usual felicity" and that the original decision had been too restrictive on the exercise of the doctors' "normal clinical discretion" by specifically saying it was unnecessary to treat Baby C with antibiotics in the event of infection or set up intravenous infusions or nasogastric feeding regimes. Baby C should be treated "in a manner appropriate to her condition," and just what is appropriate should be left to her doctors to decide at the time. On the evidence submitted, the court conceded that it would be in the child's best interests to ease her suffering, pain, and distress "rather than to achieve a short prolongation of her life." As one commentator put it, "commonsense prevails . . . and clinical freedom of action is preserved."

From these and other cases heard in British courts, it seems unlikely that there will be any British legislation to deal directly with these dilemmas in the near future. The law, which grants protection to any child born alive, is unchanged, and it is always possible for a doctor to face a serious criminal charge after any decision to allow an infant to die. It is unlikely, however, that a doctor who acts in good faith and in accordance with accepted medical practice would be found guilty of murder.

Down syndrome provides a good example of how the "ethical climate" has changed during the relatively short time that has elapsed since these issues became matters of public debate in the early 1970s. Unlike Trisomy 13 or Trisomy 18 where the outlook for health and life is uniformly bleak, children with Down syndrome may have a wide spectrum of abnormalities and intelligence, and given good care, they may survive to a relatively old age. Most require special education and remain dependent, but an increasing number, given the appropriate stimulation and support, have demonstrated their ability to start mainstream education. It is ridiculous to suggest that they have an unacceptably poor quality of life if they grow up within a loving family. Thus, as endorsed by court decisions on both sides of the Atlantic, these infants are treated no differently from infants without a chromosome abnormality. Ethical and legal difficulties will still arise when the infant is rejected by the parents or if they refuse permission for a life-saving procedure, such as an operation to correct intestinal obstruction, a common complication of Down syndrome.

In my experience, lasting rejection of the infant is unusual, but occasionally parents may reject a life-saving operation because they believe sincerely that it is not in their infant's best interests. Whatever cynics might say, inconvenience, costs, or the effect on them or others in the family are rarely uppermost in the parents' minds. In the past, I have supported parents who refused permission for surgery. After Baby Doe, such a decision would be unthinkable in the US, and following the pattern of recent cases in the UK, it is most unlikely that the courts would find in favor of the parents with or without the support of the doctors (Williams, 1981).

Some lawyers continue to be unhappy about a lack of clarity in the law as it affects abnormal or extremely immature infants. Others warn of the dangers of creating legislation that almost inevitably would be too restrictive—too blunt an instrument to deal with sensitive and tragic episodes that affect different families very differently. Some latitude and tolerance in decision making must be expected and given. One prominent legal scholar has been particularly critical of what he sees as the doctors' unwillingness to lay down rules of conduct that could be approved by the courts.

It is a matter of continuing astonishment to me that doctors resist the notion that there can be rules or guidelines stipulating how they ought to act. For again, whether or not it is realized or admitted, doctors are already invoking them when making their decisions. They do not decide in some ethical or legal vacuum (Kennedy, 1988).

Although decisions about life-sustaining treatment in the UK are still left largely to the discretion of the doctors and parents, guidelines on which to base these decisions are gradually emerging through a combination of these "test cases," and various publications by pediatricians who are willing to submit their policies and practices to open debate (Campbell, 1982; Whitelaw, 1986; Walker, 1988).

Current Trends

The 1973 report from an American neonatal unit indicated that 14% of the deaths over a two-and-a-half-year period had followed the withdrawal of life-prolonging treatment. Surveys in a similar neonatal unit in Scotland show that the proportion of neonatal deaths that follow such decisions is increasing, and it is interesting to speculate if it would have been prudent or even possible to report such data from a hospital in the United States after Baby Doe (Campbell et al., 1988). It might be concluded that intensive treatments, such as Intermittent Positive Pressure Ventilation (IPPV), are being used even more selectively, and that more and more babies are being denied intensive care and allowed to die, but in the US and Europe, infant mortality rates have fallen equivalently. With a reduction in the absolute number of deaths, it is only to be expected that, if life-sustaining treatment is used responsibly, there will be a proportionate increase in the deaths that follow the withdrawal of treatment. In the same way, deaths from withholding certain treatments will increase proportionately as more and more infants are identified as live-born at the limits of viability

(22–24 weeks' gestation). Thus, as mortality falls, it seems inevitable and proper that the proportion of the deaths attributable to "nontreatment decisions" will increase as the futility and burdensome nature of further treatment becomes all too apparent to parents and staff. Persisting with life-sustaining treatment for some infants is not a matter of saving life, but of prolonging the process of dying. "Drawing lines" at gestational ages when there is considerable risk of neurological disability in the survivors or when there is evidence of significant brain damage seems entirely appropriate and ethical.

Does this practice reflect what is happening in the rest of the UK? We think it does, although accumulated experience and resources will dictate some local variation in the criteria for intensive treatment, particularly of the VLBW infants. Recent reports suggest that most pediatricians will treat, at least for a time, almost all infants born alive to give them "a trial of life" until further assessment and discussion with the family can clarify the diagnosis, prognosis, and the family's views on the options available and the likely outcomes with or without treatment. It is felt to be particularly important for clear guidelines to be available to junior staff who usually attend these infants at birth. Our instructions are that all infants of whatever weight or gestational age should be resuscitated, unless a decision to the contrary has been reached with the parents prior to delivery, e.g., following the detection prenatally by scanning of a major brain defect like anencephaly.

For VLBW infants, some units operate a flexible "cutoff" below which assisted ventilation and other life-supports would not be continued *routinely* without at least establishing that there was no evidence of significant brain hemorrhage. Other units, especially if well staffed and equipped, will treat all infants aggressively perhaps for months and will withdraw care only very reluctantly. Some, particularly in teaching hospitals, defend this practice by insisting that it is their mission to extend the frontiers of neonatal intensive care. From the challenges of keeping alive tinier and tinier infants, they hope to derive new knowledge, skills, and techniques that may eventually improve the care given to all infants.

Transatlantic Contrasts

Withholding or Withdrawing

Studies have compared treatment and nontreatment policies in the United States and some European countries, notably England, Scotland, and Sweden (Young, 1984; Rhoden, 1986). For example, the late Nancy Rhoden, who contributed so much with her lucid and commonsense interpretations of these medical, ethical, and legal intersections, found that doctors in Sweden seemed more willing to withhold treatment, but were reluctant to withdraw it once started, whereas in the US, generally there was a more aggressive initiation of treatment for almost all viable infants with intensive care continued until death seemed certain, "the wait until certainty approach."

Rhoden preferred the British compromise, "the individualized prognostic strategy," in which treatment was initiated for most infants, but British pediatricians seemed more willing to withdraw treatment when further information and subsequent developments indicated that death or severe brain damage was a likely outcome. Many doctors see a major difference between withholding and withdrawing treatment, but I see no moral distinction whatsoever. If IPPV will not benefit a patient, not starting it or stopping it are of equal moral weight. Put another way, if IPPV will benefit the patient, it is equally wrong to withhold *or* withdraw it. If there *is* any difference, surely withdrawing treatment is preferable, since the doctors and the parents will have had more time to assemble the facts, to reflect on the issues, and to discuss the options before taking such a fundamental step. In other words, withdrawal decisions are likely to be made on better facts and firmer grounds than earlier and perhaps premature decisions to withhold treatment. It is essential to reduce to a minimum the degree of uncertainty that is always present.

Confusion about the ethics (or legality) of withholding and withdrawing treatment may also influence decision making in a subtle, but potentially dangerous way. If it is psychologically or medico-legally more difficult to withdraw treatment, doctors may withhold treatment simply out of concern that, once treatment has started, it may become much more difficult or impossible to stop. Thus, even in circumstances of considerable uncertainty, e.g., at the birth of a VLBW infant, IPPV may be denied when it could be to the infant's considerable benefit.

Uncertainty

Rhoden also points out that the preferred strategy allows for a wide variation in treatment decisions with doctors differing as to "how likely death or devastating impairment must be before withdrawal of treatment becomes a legitimate option," and that the real difference between American and British practices is the degree of certainty they seek. It is precisely because of this dilemma that I believe the views of parents and their family circumstances must be given consideration. As Rhoden indicates, it is an important question to raise. "The Swedes virtually never raise the question. The British almost always did. American doctors are acutely aware of it although many say the current legal climate has inhibited forthright confrontation of this issue."

Medical Paternalism

It is important to recognize another current transatlantic difference. "Medical paternalism" long renounced in the United States (at least by ethicists) still plays some part in European medical practice, sometimes in spite of a doctor's best efforts to avoid it. A doctor may take all possible steps to grant his or her patients (or their parents) autonomy in decision making, but compared to the US, in my experience Scottish parents usually do not wish to "dig deep" into the facts or opinions, the risks or the benefits, or the options for treatment, although these and other issues will be raised in discussion. They may express their views and seek time for reflection and further discussion, perhaps with other family members or the family doctor and clergyman. In most circumstances, they will indicate that they wish the doctor to do what seems best for their baby. The duty for making the final decision usually rests with the responsible doctor, who is expected and trusted to know the facts and to have done all the necessary homework.

Attitudes and Expectations

Some other factors also contribute to these transatlantic differences in approach. Possibly everywhere, but particularly in the United States in recent years, the image of the doctor has become tarnished with a loss of trust between doctors and patients. Some of this change in attitude may have resulted from the public's perception of doctors as high earners linked to the profit motive in providing medical care, features that, so far, have been absent from the British National Health Service. Perhaps, too, the American public, with its higher expectations and its greater enchantment with technology, feels a correspondingly greater sense of let down and betrayal if the results are imperfect. There seems little doubt that the fear of legal liability plays a bigger role in American than in European society (Huber, 1988). The prolife lobby appears to be much more evident in the US, where its members seem prepared to go to any lengths, even violence, to impose their views on others.

Ethics Committees

A useful and potentially valuable development of the Baby Doe controversy was the establishment of infant care review committees (Annas, 1984). If appropriately constituted, they can provide a broader forum for the discussion of the complex issues that surround the care of these infants, and help the primarily responsible doctors to come to a decision that is soundly based on ethics and law. They can also assist in the formation of hospital policies and practices that will conform to the requirements of the states' Child Protective Services Agencies, and thereby provide some ethical comfort and legal protection to the doctors, who feel vulnerable to crusading moralists or lobbyists for various causes (Fleischman, 1986).

In the UK at present, such committees do not exist, at least not formally, although many individuals may meet to discuss difficult dilemmas. (British Hospital Ethics Advisory Committees are equivalent to the American Institutional Review Boards and deal almost exclusively with the ethics of research.) I think that British pediatricians would welcome the advice and support of an Infant Care Review Committee, provided that the committee was not authorized or expected to make the final decision, a danger, it seems, that some American committees have not managed to avoid. If that happens, doctors, for obvious reasons, find it advisable or perhaps easier to "pass the buck" of medical decision making to the committee or a small core of its members. The committee, in turn, may fail to agree or may postpone making a decision even when one is desperately needed. Being somewhat remote from the reality of the intensive care nursery, some committee members may also pay less regard than they should to the needs of the individual infant and the family. The committee may be unduly influenced by one or more of its members to make decisions that protect other interests or, as in the case discussed earlier, those of the institution.

If the doctor primarily responsible for the care of the infant abdicates his responsibility for decision making to a committee, the lives of some infants will be prolonged. It is certainly possible that the occasional infant may be saved from an untimely death by preventing the inappropriate withdrawal of treatment. However, unless doctors can no longer be trusted to make the "right" decisions most of the time, committee prevarication will merely result in the pointless prolongation of dying for many more infants at great cost to infants, families, and society. This avoidance of responsibility also devalues the role of a doctor in providing leadership in medical care, a trend that must be to the detriment of the profession and ultimately to society (Ingelfinger, 1973).

Economic Costs

Apart from concerning ourselves with the medical, ethical, and personal realities of intensive care for individual patients, should we be asking questions like: Can we continue to provide intensive care indiscriminately for all without regard to the costs and the economic consequences for others? Should cost concern a doctor whose first duty is to his or her patient? Should doctors be using power over life and death more responsibly by limiting it to those who can benefit the most? In other words, should we ration intensive care? If the answer to the last question is yes, this begs some others: What criteria should be used? Who should be the gatekeepers? Can doctors continue to select treatment on medical and ethical criteria to an extent that will retain public confidence and trust, to say nothing of the law? By so doing, can we perhaps avoid or at least postpone the imposition of rationing based primarily on economic considerations and determined by others who will not necessarily have the interests of individual patients at heart?

Harris, referring to the United States, wrote:

> ... Our basic national problem is that we physicians refuse to grapple with a limited resource economy. We want to provide medical care with no cost restraints. Whatever number of dollars it takes to provide care to indigents—or to perform transplantation, dialysis, immunizations, etc. ... must be provided ... if we as physicians do not accept the responsibility of how best to use our resources, others of necessity, will (Harris, 1988).

This is happening now. I read recently that:

> Oregon has set priority care standards emphasizing preventive medicine, pre-natal care and acute care.

Transplants are not paid for and desperate patients and children have done painful television appeals for money to pay for an organ transplant.

Although health spending in the US is approximately twice what it is in the UK, several surveys have shown considerable discordance in the way the respective health-care systems are viewed by the public. There are, of course, major cultural differences in attitudes to health and health care that must be acknowledged, but despite our well-publicized problems, several surveys in the UK still reveal a remarkable degree of public and patient satisfaction with the performance of the National Health Service, and this is further reflected in widespread public concern about recent government proposals for change. On the other hand, as reported in the *New York Times*, a recent survey showed that 89% of Americans believe that their system needs fundamental change. "We Americans have the most expensive, the least well liked, the least equitable, and in most ways the most inefficient systems."

Some seem not too concerned about these escalating costs. Professor Mark Pauly, of the University of Pennsylvania, when attending a conference at the Institute of Economic Affairs in London, was quoted as being "not bothered" by an annual rise in costs of 6%, since by definition it was the right rate because it was set by the market. However, much of this growth is accounted for, not by an increase in the volume of services, but by higher payments for procedures and higher intensities of treatment. It provides no solution to the problem of equity. "The poor remain a problem in the free market" (Smith, 1988).

What it all means is that, in either system, some form of rationing is inevitable. How to ration equitably will continue to be the subject of much agonized debate in the 1990s, but it is worth remembering that rationing of one form or another has always been a fact of life. Time and energy devoted to one patient are of necessity denied to another, and it has always been necessary for doctors to make choices between different patients, usually with the best motives.

Since the introduction of our National Health Service in 1948, general practitioners have acted as "gatekeepers" to determine access to certain treatments and services. It is well known that kidney dialysis, intensive care, and coronary artery surgery, to mention only a few, are much less widely available in the UK than they are in the US. Developments in high-technology medicine are more expensive, and the implications of denial to certain patients are much more serious. They provide fuel for media exploitation, which is dramatic and newsworthy. Thus, the choices to be made are that much more difficult. In the UK to date, rightly or wrongly, much of this paternalistic rationing has been "fudged" or "rationalized," justified ethically as being in the best interests of the patients and/or by the equitable distribution of scarce resources. In many situations, the criteria remain crude, medically unsatisfactory, ethically suspect, and not explicit enough for others to determine easily if justice is being done.

Neonatal intensive care certainly is expensive, but is it money well spent? A study from Canada indicated that, by every economic measure, neonatal intensive care for infants weighing 1000–1499 grams is superior to neonatal intensive care for infants weighing 500–999 grams (Boyle et al., 1983). As the authors emphasized, this study was an economic one that considered both outcomes and costs.

> A very different conclusion could result if only clinical outcomes were considered, for example neonatal intensive care of infants weighing 750–999g in the largest gain in survival for any sub-group [19 to 43%], although this sub-group also produced the largest net economic loss.

A difficulty with this Canadian study and others that purport to show cost effectiveness in strictly economic terms is that infants cannot be packaged neatly into birthweight bundles and rated accordingly. To reduce such recommendations to absurdity, we could argue that, if an infant weighed 998 grams at birth, intensive care might be withheld, since it would be devoted to one weighing 1001 grams. The grouping of birthweight or gestational age categories is useful in comparative statistics, but such selection is arbitrary, the averaging of cost data from an intensive care unit is extremely complex, and comparisons among individual centers to date have been frustrated because of differences in infant selection, costing guidelines, and levels of care.

In the UK, studies from Liverpool and Birmingham showed striking differences in total costs per infant and differences between survivors and nonsurvivors. The cost per survivor in Birmingham was almost six times that for a nonsurvivor (Newns et al., 1984), whereas in Liverpool, a survivor cost only 23% more than a nonsurvivor (Sandhu et al., 1986). These cost differences could be the result of regional differences in the proportion of infants transferred after birth or *in utero*, but could also be the result of differences in the duration and intensity of effort put into the management of extremely sick infants with a poor prognosis, *such efforts being determined by professional and parental attitudes.* There is nothing necessarily wrong with this, because it suggests that decisions were focused firmly on what seemed best for individual infants without regard to the economic costs of care.

These and other studies on cost-effectiveness raise a number of ethical questions. For example, should countries or regions with insufficient capacity to provide expensive care for all low-birthweight infants give priority to infants over 1000 grams? When should a line be drawn? Should we not be paying more attention and spending more of our resources in seeking alternatives, and perhaps more logical approaches, to solving the problem of premature birth? Providing better antenatal care for young and socially disadvantaged mothers, and attempting improvements in their health, nutrition, and housing might be more effective, as seems to be the case in Scandinavia. These measures require major changes in public and political attitudes, and the appropriate social and perhaps legislative action, all of which take time. They are also less immediately appealing than increasing the funds for intensive care.

At present, the microallocation of high technology resources within each neonatal unit is largely determined by what the intensive care team of doctors and nurses believe to be best for each infant. To me, using intensive care selectively in this way, with due recognition, albeit limited, of parental autonomy, seems more defensible morally than deciding treatment on criteria based on the interests of others or on the economic costs to the unit, hospital, funding agency, or the state. At the same time, using expensive treatment only when it will benefit the infant must also make economic sense.

The Law

In 1973, we ended our paper with these words: "If working out this dilemma in ways such as these we suggest is in violation of the law, we believe the law should be changed." Perhaps we were thinking of something akin to the Limitation of Treatment Bill proposed for the UK, but which, as discussed earlier, could erode some of the trust between doctors and patients that is so important to retain. On reflection, such legislation would probably be unwise. There is some protection against inappropriate, misguided, or ignorant decision making in an unchanged law with doctors liable to legal scrutiny if abuses are suspected. Doctors must be careful not to impose unwelcome choices on parents or create conflicts within families through a sense of moral superiority, yet they may have to guide parents toward what is acceptable not only medically, but morally, socially, and within the law.

Occasionally parents' understanding of the medical realities and their views on the options available may be so faulty that disagreement will result, and it will be necessary to seek court assistance in order to proceed with urgent treatment or even, very occasionally, to permit withdrawal of pointless treatment (Paris et al., 1990). Sometimes parents will ask a doctor to "do something" to hasten the death of a loved one whose prolonged dying is causing great anguish. Although such an action might be viewed as morally correct, the parents must understand that it would be against the law, and there could be legal consequences for them as well as for the doctor.

In a recent review of the law regarding "selective nontreatment," a British professor of forensic medicine and an American lawyer offered a "view from the mid-Atlantic," but their guidelines for the exercise of parental choice seem much closer to the shores of Europe.

> The parents who have conceived the infant and who have responsibility to raise it, should be given the right, within closely and carefully drawn confines, to elect non-treatment when their child is born severely deformed. Additionally, society should not, without strong reason, dictate standards to physicians, which compel treatment in circumstances in which many ethically minded doctors would feel it was medically inappropriate (Mason and Meyers, 1986).

Conclusion

One aim of our 1973 report was to "break public and professional silence on a major social taboo and some common practices." This was achieved beyond all expectations, but we cannot claim equal success for the hope that "out of the ensuing dialogue perhaps better choices for patient and families can be made." Following Baby Doe, families or their pediatricians now appear to have little choice or discretion in deciding treatment, and much of their autonomy in decision making has become subjugated to an ethics committee or the State Child Protection Agency. The evidence that the choices are "better" for parents or families is not reassuring. Rather, the opposite is true.

We sought to open debate on these issues because we had become concerned that high-technology medicine, its excitement and triumphs notwithstanding, increasingly was being applied, relentlessly and indiscriminately, to newborn babies without due regard to the adverse consequences for some infants and some families. We believed that doctors were paying too little attention to the dilemmas of "human ambiguity," too little attention to parental views and their hopes for the future, too little attention to pain, suffering, and future quality of life, and too little attention to the realities of family living in individual circumstances. Recent developments have increased that concern.

References

Annas, G. J. (1984) Ethics committees in neonatal care: Substantive protection or procedural diversion? *Am. J. Public Health* **74,** 843–845.

Boyle, M. H., Torrance, G. W., Sinclair, J. C., and Horwood, S. P. (1983) Economic evaluation of neonatal intensive care of very-low-birthweight infants. *N. Engl. J. Med.* **308,** 1330–1337.

Brahams, D. (1989) Medicine and the law: Court of appeal endorses medical decision to allow baby to die. *Lancet* **i,** 969–970.

Brahams, D. and Brahams, M. (1983) The Arthur case: A proposal for legislation. *J. Med. Ethics* **9,** 12–15.

Campbell, A. G. M. (1982) Which infants should not receive intensive care? *Arch. Dis. Child.* **57,** 569–571.

Campbell, A. G. M., Lloyd, D. J., and Duffty, P. (1988) Treatment dilemmas in neonatal care: Who should survive and who should decide? *Ann. NY Acad. Sci.* **447,** 19–28

Duff, R. S. and Campbell, A. G. M. (1973) Moral and ethical dilemmas in the special care nursery. *N. Engl. J. Med.* **289,** 890–894.

Duff, R. S. and Campbell, A. G. M. (1980) Moral and ethical dilemmas: Seven years into the debate about human ambiguity. *Ann. Am. Acad. Polit. Soc. Sci.* **447,** 19–28.

Engelhardt, H. T., Jr. (1989) Comments on the recommendations regarding Section 504 of the Rehabilitation Act of 1973 and the Child Abuse Amendments of 1984, in *US Commission on Civil Rights: Medical Discrimination Against Children with Disabilities.* **447,** 158–165.

Feinberg, W. M. and Ferry, P. C. (1984) A fate worse than death: The persistent vegetative state in childhood. *Am. J. Dis. Child.* **138**, 128–130.

Fischer, A. F. and Stevenson, D. K. (1987) The consequences of uncertainty: An empirical approach to medical decision-making in neonatal intensive care. *JAMA* **258**, 1929–1931.

Fleischman, A. R. (1986) An infant bioethical review committee in an urban medical center. *Hastings Cent. Rep.* **16, 3,** 16–18.

Gunn, M. J. and Smith, J. C. (1985) Arthur's case and the right to life of a Down's syndrome child. *Criminal Law Rev.* 705–715.

Hack, M. and Fanaroff, A. A. (1989) Outcomes of extremely-low-birthweight infants between 1982 and 1988. *N. Engl. J. Med.* **321**, 1642–1647.

Harris, C. J. (1988) Gate-keepers and cost containers in HMOs. *N. Engl. J. Med.* **318,** 1698.

Harvey, D., Cooke, R. W. I., and Levitt, G. A. (1989) *The Baby under 1000g* (Wright, London and Boston).

Huber, P. W. (1988) *Liability: The Legal Revolution and Its Consequences.* (Basic Books Inc., New York).

Ingelfinger, F. J. (1973) Bedside ethics for the hopeless case. *N. Engl. J. Med.* **289**, 914,915.

Jennett, B. (1984) *High Technology Medicine* (Nuffield Provincial Hospitals Trust, Oxford).

Kennedy, I. (1988) *Treat Me Right: Essays in Medical Law and Ethics* (Clarendon Press, Oxford), pp. 154–174.

Koop, C. E. (1978) The sanctity of life. *J. Med. Soc. NJ* **75**, 62–69.

Kopelman, L. M., Irons, T. G., and Kopelman, A. E. (1988) Neonatologists Judge the "Baby Doe" regulations. *N. Engl. J. Med.* **318,** 677–683.

Lantos, J. (1987) Baby Doe five years later: Implications for child health. *N. Engl. J. Med.* **317,** 444–447.

Lyon, J. (1985) *Playing God in the Nursery* (W. W. Norton and Co., New York and London).

Mason, J. K., and Meyers, D. W. (1986) Parental choice and selective nontreatment of deformed newborns: A view from mid-Atlantic. *J. Med. Ethics* **12,** 67–71.

Moskop, J. C. and Saldanha, R. L. (1986) The Baby Doe rule: Still a threat. *Hastings Cent. Rep.* **16, 2,** 8–14.

Newns, B., Drummond, M. F., Durbin, G. M., and Culley, P. (1984) Costs and outcomes in a regional neonatal intensive care unit. *Arch. Dis. Child.* **59**, 1064–1067.

Paris, J. J., Crone, R. K., and Reardon, F. (1990) Occasional notes: Physicians' refusal of requested treatment. The case of Baby L. *N. Engl. J. Med.* **322**, 1012–1014.

President's Commission for the Study of Ethical Problems in Medicine and Biomedical and Behavioral Research (1983) *Deciding to Forego Life-Sustaining Treatment: Ethical, Medical and Legal Issues in Treatment Decisions* (US Government Printing Office, Washington, DC).

Ramsey, P. (1978) *Ethics at the Edges of Life: Medical and Legal Intersections* (Yale University Press, New York and London).

Rhoden, N. K. (1986) Treating Baby Doe: The ethics of uncertainty. *Hastings Cent. Rep.* **16(4)**, 34–42.

Sandhu, B., Stevenson, R. C., Cooke, R. W. I., and Pharoah, P. O. D. (1986) Cost of neonatal intensive care for very-low-birthweight infants. *Lancet* **i**, 600–603.

Shapiro, D. L. and Rosenberg, P. (1984) The effect of federal regulations regarding handicapped newborns: A case report. *JAMA* **252**, 2031–2033.

Smith, J. (1988) Little to learn from the Americans. *Br. Med. J.* **296**, 1280.

US Commission on Civil Rights (1989) *Medical Discrimination against Children with Disabilities* (US Government Printing Office, Washington, DC).

Van Dyke, D. C., and Allen, M. (1990) Clinical management considerations in long-term survivors with Trisomy 18. *Pediatrics* **85**, 753–759.

Walker, C. H. M. (1988) Current topic. ".....officiously to keep alive." *Arch. Dis. Child.* **63**, 560–564.

Whitelaw, A. (1986) Death as an option in neonatal intensive care. *Lancet* **2**, 328–331.

Williams, G. (1981) Down's syndrome and the duty to preserve life. *New Law J.* **Oct. 1**, 1020–1021.

Young, E. W. D. (1984) *Societal Provision for the Long-Term Needs of the Disabled in Britain and Sweden Relative to Decision-Making in Newborn Intensive Care Units* (World Rehabilitation Fund Inc., New York).

Young E. W. D. and Stevenson, D. K. (1990) Limiting treatment for extremely premature low-birth-weight infants (500–750g). *Am. J. Dis. Child.* **144**, 549–552.

Neonatologists, Pediatricians, and the Supreme Court Criticize the "Baby Doe" Regulations

Loretta M. Kopelman,
Arthur E. Kopelman,
and Thomas G. Irons

Introduction

Earlier we reported neonatologists' reaction (Kopelman et al., 1988) to the controversial federal regulations governing the treatment of severely handicapped infants—the "Baby Doe" Regulations (DHHS, 1985). We found that the responding neonatologists were highly critical of these regulations, and that their concerns were similar to those expressed by the United States Supreme Court (*Bowen v. Am. Hosp. Ass'n.*, 1986) in rejecting an earlier set of Baby Doe regulations

From: *Compelled Compassion* Eds.: Caplan, Blank, and Merrick
©1992 The Humana Press Inc.

(DHHS, 1984). This agreement among the neonatologists and legal authorities on the Supreme Court lead us to conclude that these regulations were ill-considered. Some have questioned, however, whether the neonatologists' negative reaction to these rules as reported in our survey might be biased, since the regulations restrict their daily practice. Others question if neonatologists have an unrealistic, pessimistic picture of severely sick newborns' prognoses, since they do not generally deal with older children. To help answer these questions, we now present the reaction of nonneonatologist pediatricians to these regulations. These data were collected at the same time as the reported survey. Their responses are similar to the neonatologists' responses, undercutting both objections to the conclusions of our survey.

The current Baby Doe regulations (DHHS, 1985) are based on the 1984 amendments to the Child Prevention and Treatment Act (US, 1984). An earlier version of the Baby Doe regulations (DHHS, 1984), promulgated in 1984 on the basis of Section 504 of the Rehabilitation Act of 1973 (US, 1984), was struck down by the US Supreme Count on June 9, 1986 (*Bowen v. Am. Hosp. Ass'n.*, 1986). Although the substance of the two sets of regulations is similar, the earlier regulations were issued under civil rights law, which states must follow, whereas the recent regulations apply to federal grants to the states. States can refuse the federal funds in question and, thus, be under no obligation to comply with the current regulations. (Pennsylvania, Hawaii, and California have done so.) Unlike the earlier version, the current regulations have not been fully tested in the courts. We shall call the current federal regulations BDR-II (DHHS, 1985), and the earlier set rejected by the Supreme Court, BDR-I (DHHS, 1984).

In order to evaluate the utility and influence of the current Baby Doe regulations (BDR-II), we surveyed two groups of pediatricians in September of 1986. All were active and practicing members of the American Academy of Pediatrics. First, we surveyed all members of the Perinatal Pediatric Section of the American Academy of Pediatrics. Nearly 50% of these members (we call this group "neonatologists," since almost all of them are neonatologists) responded to our questionnaire. The second group was a randomly selected sampling of the general membership prepared and sent to us by the American Academy of Pediatrics. We call this group "pediatricians." About 25% responded to our questionnaire.

In what follows, we compare the previously reported views of neonatologists to the Baby Doe-II regulations to the views of them expressed by the responding pediatricians who are not neonatologists. We will also compare their views to those expressed by the United States Supreme Court in rejecting an earlier, similar set of Baby Doe regulations (BDR-I). The method of moral reasoning that we will use is to consider what view can be supported by reasonable and informed persons of good will. The agreement of experts in the law, experts in the care of newborns, and experts in the care of older children should be a powerful indication of what is a good-willed, reasonable, and informed opinion in these matters.

Methods

On September 30, 1986, our questionnaire was mailed to all those whose names were supplied to us by the American Academy of Pediatrics. They fell into two groups. First, we requested the names of and sent the survey to all 1007 members of the Perinatal Pediatric section of the American Academy of Pediatrics. Second, we requested a random sampling of 3000 practicing pediatricians who were not neonatologists; a sample of 2996 pediatricians was prepared by the American Academy of Pediatrics. By the cutoff date, December 1, 1986, we received responses from 494 neonatologists (49%) and from 730 pediatricians who were nonneonatologists (25%). The questionnaire consisted of nine general statements about appropriate treatment in three hypothetical cases of severely handicapped newborns with life-threatening conditions. As background, we quoted the passage from current Baby Doe-II regulations that describes the exceptions to the requirement that all infants with life-threatening conditions be treated.

[The withholding of medically indicated treatment is]
the failure to respond to the infant's life-threatening
conditions by providing treatment (including appropriate nutrition, hydration, and medication) which, in
the treating physician's (or physicians') reasonable
medical judgment, will be most likely to be effective
in ameliorating or correcting all such conditions,
except that the term does not include the failure to
provide treatment (other than appropriate nutrition,
hydration, or medication) to an infant when, in the
treating physician's (or physicians') reasonable medical judgment any of the following circumstances
apply: (i) The infant is chronically and irreversibly
comatose, (ii) The provision of such treatment would
merely prolong dying, not be effective in ameliorating
or correcting all of the infant's life-threatening conditions, or otherwise be futile in terms of the survival of
the infant, or (iii) the provision of such treatment would
be virtually futile in terms of the survival of the
infant and the treatment itself under such circumstances would be inhumane. (DHHS, 1985)

At the end of the questionnaire, a space was provided for
comments.

The three case histories were chosen to illustrate current (1986) ethical dilemmas faced by neonatologists caring
for infants with life-threatening conditions and with
extremely poor chances for survival and for cognitive development. Treatment could only be expected to extend their
lives for a relatively short time. The three cases were created to match situations in which major textbooks recommend only comfort care as appropriate (Avery, 1987; Smith,
1982). The problem for respondents was to decide whether
maximal life-sustaining treatment or comfort measures were
in the infant's best interest and what would be required by
current federal regulations. These three cases are still good
examples for these purposes. Six responses to each statement
were possible: strongly agree, agree, uncertain, disagree,
strongly disagree, not applicable. One item was omitted

because it was ambiguous. The "not applicable" and "no response" categories were dropped because few checked them. The neonatologists rarely checked them (range 0–3%, average 1%). The pediatricians selected them infrequently (range >1–6%, average 2.2%). The results presented herein combine the "Strongly Agree–Agree" and "Strongly Disagree–Disagree" responses.

Study Population

We received responses from 494 neonatologists (49%) among the 1007 who received questionnaires in our single mailing. Returns came from neonatologists in 47 states, the District of Columbia, and Puerto Rico. The mean number of responses from each state or district was 9.3 (range 1–47). Most respondents (77%) were male; 14% received their MD degrees before 1961, and 55% after 1970. We have presented their combined responses.

We received responses from 730 pediatricians who were nonneonatologists (25%) among the 2996 who received questionnaires in our single mailing. Returns came from pediatricians in 50 states, the District of Columbia, and Puerto Rico. The mean number of responses from each state or district was 14.0 (range 1–135). Most respondents (78%) were male; 43% got their MD degrees before 1961, and 20% after 1970. We have presented their combined responses.

Results

Responses to Cases

The results are presented first for neonatologists and then for other pediatricians. In responding to statements about *Case 1*—that of a full-term infant with Trisomy 13 who has congestive heart failure at three weeks of age (Table 1)— 77% of the neonatologists responded that the parents' wishes would influence their decision about whether to perform cardiac catheterization, a diagnostic study that might lead to

Table 1
Case 1 (1986)[a]

A term infant with Trisomy 13 Syndrome develops congestive heart failure at 3 weeks of age:

1. Good medical judgment requires a cardiac catheterization be performed after anticongestive treatment is begun

	% Agree	% Uncertain	% Disagree
Neonatologists	10	4	86
Other pediatricians	27	10	63

2. The BDR-II require that cardiac catheterization be performed

Neonatologists	22	17	61
Other pediatricians	44	23	32

3. The parents' wishes would influence my decision

Neonatologists	77	8	15
Other pediatricians	80	8	12

---

[a]Owing to rounding, some add up to 99 or 101%.

cardiac surgery; 8% were uncertain; and 15% said that the parents' wishes would not influence the decision. The majority, 61%, agreed that the BDR-II did not require this lifesaving treatment; 22% stated that the BDR-II required cardiac catheterization; 17% were uncertain. Most (86%) held that good medical judgment did not require that a cardiac catheterization be performed after anticongestive treatment was begun; 10% disagreed, and 4% were uncertain. There was a significant difference ($\chi^2 = 81$, df = 2, P = <0.001) between the responses about what was required by good medical judgment and by the BDR-II (Zar, 1984). Of the 415 neonatologists (86%) who believed that good medical judgment did not require them to do a cardiac catheterization, 17% (68 out of 415) believed that the BDR-II required that cardiac catheterization be performed.

Pediatricians, in responding to this hypothetical case about an infant with Trisomy 13, *Case 1* (Table 1), reacted as follows: 80% responded that the parents' wishes would influence their decision about whether to provide heroic care; 8% were uncertain; 12% disagreed. There was little agreement about whether or not the BDR-II required this life-saving treatment: 44% thought that they required that a cardiac catheterization be done, 32% that they did not, and 23% were uncertain about whether it was required. The majority of pediatricians (63%) held that good medical judgment did not require that a cardiac catheterization be performed after anticongestive treatment was begun; 27% disagreed; 10% were uncertain about this. As with neonatologists, there was a significant difference (χ^2 = 89.5, df = 2, P = <0.001) between the responses about what good medical judgment and what the BDR-II require. Of the 437 pediatricians (63%) who believed that good medical judgment did not require them to do a cardiac catheterization, 20% (139 out of 693) believed that the BDR-II required that cardiac catheterization be performed.

In responding to statements about *Case 2*—that of an extremely premature infant with a cerebral hemorrhage (Table 2)—75% of the neonatologists said that they would consider stopping ventilator support when the infant's condition deteriorated, but 16% said they would not consider such a step, and 9% were uncertain. Over four-fifths (87%) said the parents' wishes would influence their decision; 8% said they would not influence the decision; 5% said they were uncertain. There was considerable disagreement about whether the BDR-II required continuation of ventilator support in such a case: 30% thought they did; 52% thought they did not; 18% were uncertain. Over two-thirds of respondents (68%) reported that their approach to the care of such infants had not changed as a result of the BDR-II, but 23% said that it had in such cases, and 9% were uncertain. Over four-fifths (86%) disagreed with the statement that their treatment of such an infant would be based on medical facts alone and would not take into account the parents' wishes; 8% agreed with the statement, and 6% were uncertain. Many saw a conflict between what they believed was required by

Table 2
Case 2 (1986)[a]

You are present at the delivery of a 550-gram premature infant
who is born at 25 weeks' gestation. There is no evidence of
asphyxia. In your hospital, 10% of such infants survive. At one
day of age, the same infant's anterior fontanel is bulging, there
is anemia, and the infant is flaccid except for seizure activity.
Vital signs are stable. Cranial ultrasound demonstrates a large
cerebral intraparenchymal hemorrhage. In your hospital, no
more than 5% of such infants survive, and almost all survivors
have severe neurological handicaps.

1. I would consider stopping ventilator support at this time

	% Agree	% Uncertain	% Disagree
Neonatologists	75	9	16
Other pediatricians	71	11	18

2. The parents' wishes would influence my decision:

Neonatologists	87	5	8
Other pediatricians	83	6	11

3. The BDR-II requires me to continue ventilator support

Neonatologists	30	18	52
Other pediatricians	40	27	33

4. My decision would be based on medical facts, and the parents'
wishes would not enter into it

Neonatologists	8	6	86
Other pediatricians	14	8	78

5. My approach to the care of such infants has changed as a
result of passage of the BDR-II

Neonatologists	23	9	68
Other pediatricians	36	17	48

[a]Owing to rounding, some add up to 99 or 101%.

the BDR-II and what they thought was the best medical
choice for the infant. There was a significant difference (χ^2 =
186, df = 2, P = <0.001) between the responses about what
they considered medically appropriate and what the BDR-II
required. Of the 353 neonatologists (75%) who said they would
consider stopping ventilator support, 23% believed that the

BDR-II required them to continue it; 59% disagreed; 18% were uncertain.

In responding to statements about *Case 2* (Table 2), 71% of the pediatricians who were nonneonatologists said that they would consider stopping ventilator support when the infant's condition deteriorated, but 18% said they would not, and 11% were uncertain. Most (83%) said the parents' wishes would influence their decision; 11% said they would not; 6% said they were uncertain.

The pediatricians were similarly divided over whether the BDR-II required continuation of ventilator support in such a case. Of the pediatricians who responded, 40% thought they did; 33% thought they did not; 27% were uncertain. They were also divided about whether their approach to the care of such infants has changed as a result of the BDR-II rules: 48% denied that their approach to the care of such infants had changed as a result of the BDR-II, 36% said that it had, and 17% were uncertain. Most (78%) said that a decision about what ought to be done for such an infant would not be based on medical facts alone, but would take into account the parents' wishes; 14% said they would decide what treatment was best without consulting the parents' wishes; 8% were uncertain. Many saw a conflict between what they believed was required by the BDR-II and what they thought was the best medical choice for the infant. There was a significant difference (χ^2 = 31.3, df = 2, P = <0.001) between the responses about what was considered in the infant's best interest and what the BDR-II required. Of the 673 (71%) pediatricians who said they would consider stopping ventilator support, 34% believed that the BDR-II required them to continue it, 39% disagreed, and 27% were uncertain.

In responding to *Case 3*—that of a full-term infant with congenital hydrocephalus, in deep stupor or coma, in whom ventriculitis is diagnosed, and whose parents do not want treatment (*see* Table 3)—77% of the neonatologists disagreed that the life-saving treatment was in the infant's best interest; 9% said it was; 14% were uncertain. Almost half (47%) of neonatologists, however, stated the BDR-II required

Table 3
Case 3 (1986)[a]

A full-term infant is born with advanced congenital hydrocephalus, is blind, and is minimally responsive (comatose). A shunt
is inserted, but becomes infected and (life-threatening) ventriculitis has just been diagnosed. The parents do not want the
infant to be treated.

1. According to the Baby Doe-II, I am compelled to treat

	% Agree	% Uncertain	% Disagree
Neonatologists	47	14	39
Other pediatricians	53	18	29

2. It is in the infant's best interest to treat

Neonatologists	9	13	77
Other pediatricians	11	11	78

3. My approach to treatment of such an infant has changed as a
   result of the BDR-II

Neonatologists	33	9	58
Other pediatricians	38	13	49

[a]Owing to rounding, some add up to 99 or 101%.

treatment for this infant; 39% believed they did not; 14% were
uncertain. A third of the neonatologists, 33%, stated that their
approach to the treatment of such an infant had changed as
a result of the BDR-II; 58% said that it had not changed; 9%
were uncertain. Thus, once again many perceived a conflict
between what they regarded to be in the infant's best interest and the requirements of the BDR-II. As in the other two
cases, the groups' responses to treatment decisions made in
the infant's best interest were significantly different than
what they said the BDR-II required ($\chi^2 = 186$, df $= 2$, P $=$
<0.001). When comparing subgroups, we found that 32% of
the neonatologists who thought that it was not in the infant's
best interest to receive life-saving treatment believed that
such treatment was required by the BDR-II.

Of the pediatricians who were not neonatologists
responding to questions about *Case 3*, 78% (*see* Table 3) dis-

agreed that the life-saving treatment was in the infant's best interest; 11% said it was; 11% were uncertain. About half (53%) of the pediatricians who were not neonatologists, however, stated that the BDR-II required treatment for this infant; 29% believed they did not; 18% were uncertain. Over a third, 38%, stated that their approach to the treatment of such an infant had changed as a result of the BDR-II; 45% said that it had not changed; 13% were uncertain. Many perceived a conflict between what they regarded to be in the infant's best interest and the requirements of the BDR-II. Unlike the other two cases, however, the nonneonatologist pediatricians' responses to questions about treatment being in the infant's best interest were not significantly different from what they said the BDR-II required (χ^2 = 2.3, df = 2, P = <0.12) at the 95% level of confidence.

Other Survey Questions (Table 4)

Most neonatologists (81%) denied that the BDR-II would result in improved care for all infants; 14% were uncertain about this; 5% thought that they would. Over three-quarters (76%) disagreed that the BDR-II were needed to protect the rights of handicapped infants; 10% were uncertain about this; 14% held that the BDR-II were needed. Two-thirds (66%) said that the BDR-II affect parental rights to consent to or refuse treatment based upon what they regarded as in the infant's best interest; 15% were uncertain; 19% believed the BDR-II would not interfere with parental rights. Many (60%) judged that the BDR-II do not allow adequate consideration of the infant's suffering; 11% were uncertain; 29% said that they did.

Many neonatologists (56%) held that critically ill infants were overtreated when the chances for their survival were very poor; 13% were uncertain about this; 31% disagreed. Over four-fifths (82%) held that, if the federal government required intensive life-saving treatment of such severely ill and impaired infants, then it should guarantee payment for costs of that treatment; 6% were uncertain; 12% disagreed. Similarly, 82% stated that the federal government should guarantee payment for their subsequent rehabilitative care;

Table 4

1. The BDR-II will result in improved care for all infants			
	% Agree	% Uncertain	% Disagree
Neonatologists	5	14	81
Other pediatricians	6	16	78

2. The BDR-II were needed to protect the rights of handicapped infants			
Neonatologists	14	10	76
Other pediatricians	20	12	67

3. The BDR-II will not affect parental rights to consent to or refuse treatment based upon what is in the infant's best interest			
Neonatologists	19	15	66
Other pediatricians	19	17	65

4. The BDR-II allow adequate consideration of suffering			
Neonatologists	29	11	60
Other pediatricians	30	14	56

5. Most critically ill infants are overtreated when the chances for their survival are very poor			
Neonatologists	56	13	31
Other pediatricians	62	18	19

6. If the federal government requires life-saving treatment of severely handicapped infants, then it should guarantee payment for that treatment			
Neonatologists	82	6	12
Other pediatricians	76	8	16

7. If the federal government requires life-saving treatment of severely handicapped infants, then it should guarantee payment for their rehabilitative care			
Neonatologists	82	6	12
Other pediatricians	75	9	16

8. The BDR-II have exacerbated the shortage of NICU beds			
Neonatologists	17	32	51
Other pediatricians	24	50	25

[a]Owing to rounding, some add up to 99 or 101%.

6% were uncertain; 12% disagreed. We asked if the BDR-II have exacerbated the shortage of NICU beds: 51% said they had not; 32% were uncertain about this; 17% said that they had.

Most nonneonatologist pediatricians (78%) denied that the BDR-II will result in improved care for all infants; 16% were uncertain about this; 6% thought that they would. Two-thirds (67%) held that the BDR-II were not needed to protect the rights of handicapped infants; 12% were uncertain about this; 20% held that the BDR-II were needed. Many (65%) said that the BDR-II affect parental rights to consent to or refuse treatment based upon what they regarded as being in the infant's best interest; 17% were uncertain; 19% believed the BDR-II would not interfere with parental rights. Half (56%) judged that the BDR-II did not allow adequate consideration of the infant's suffering; 14% were uncertain; 30% said that they did.

Many pediatricians (62%) held that critically ill infants are overtreated when the chances for their survival are very poor; 18% were uncertain about this; 19% disagreed. About three-quarters (76%) of the pediatricians held that, if the federal government required intensive life-saving treatment of such severely ill and impaired infants, then it should guarantee payment for costs of that treatment; 8% were uncertain; 16% disagreed. Similarly, 75% stated that the federal government should guarantee payment for their subsequent rehabilitative care; 9% were uncertain about this; 16% disagreed. We asked if the BDR-II had exacerbated the shortage of NICU beds: 24% said they had not; 50% were uncertain about this; 25% said they had.

Comments

Of the 94 (19%) neonatologists who wrote spontaneous comments, 5% expressed explicit approval of the BDR-II, whereas 49% expressed explicit disapproval. The remaining

46% of the neonatologists expressed comments not easily classified as express approval or disapproval. Of the 111 (15%) pediatricians who wrote comments, seven (6%) made positive comments; 45 pediatricians (40%) made negative remarks, and the rest (52%) were general comments not easily classified as expressing either explicit approval or disapproval.

There were frequent and similar themes in the comments of both groups. They discussed their fears of litigation, the pressures that they believed created these regulations, their encounters with these problems, the difficulty of making these choices, and the joys or hardships of taking care of profoundly disabled children. Some responded as parents or relatives of handicapped children. There were also many remarks about the importance of consultation with families, and about whether practice was now more difficult or easier as a result of the BDR-II. The BDR-II were criticized by some for ambiguity. Others questioned the soundness of the legal basis for the BDR-II and whether this intervention was a legitimate governmental role. Some said that they felt pressure to overtreat because of the BDR-II, the new technologies, and/ or the legal climate. Four people criticized the survey, either for ambiguity or for failing to clarify the law passed by Congress better.

Discussion

Most neonatologists and pediatricians who responded to our survey indicated that the BDR-II are a mistake—that they sometimes encouraged or required the overtreatment of infants. The three hypothetical case histories describe infants with little chance of survival even with heroic care. They were chosen to illustrate how (1986) neonatologists and pediatricians would approach the care of some infants with life-threatening conditions who had extremely poor prognoses, both for survival and for cognitive development. Neonatologists and pediatricians overwhelmingly agreed about what decisions were best and that comfort care was sometimes better than heroics in these cases. This perhaps is not surprising, since these judgments followed the recommendations of standard texts (Smith, 1982; Avery, 1987). There was little agreement, however, about what the BDR-II required in these very cases. The overwhelming majority of neonatologists (77–85%) and pediatricians (80–83%) wanted to consider the parents' wishes, but 21–46% of the neonatologists and 40–53% of the pediatricians thought the BDR-II required them to treat heroically. Many of the neonatologists (21–32%) and the pediatricians (36–38%) said that, as a result of the BDR-II, they had changed their care practices for such infants. When we looked at individual responses in Case 3, almost a third (32%) of the neonatologists stated both that the BDR-II required them to provide a treatment and that it was not in the best interest of the infant. In each of the three cases, there was a significant difference between how the neonatologists responded to items about what the BDR-II required, and what they wanted to consider or do from the viewpoint of these infants' best interests. When we looked at individual pediatricians' responses for Cases 1 and 2, we also found that there was a significant difference between how the neonatologists responded to items about what the BDR-II required and what they wanted to consider or do from the viewpoint of these infant's best interest. This was not true of Case 3 for the nonneonatologist pediatricians. One way to account for this is to look at the terminology used. "Minimally Responsive" has a single meaning for neonatologists, "comatose," but other health professionals use it for less serious conditions, perhaps causing different ideas about the prognosis in Case 3.

Over half of the neonatologists (55%) thought that infants with poor prognoses were overtreated in our nation's NICUs, but even more of the pediatricians (61%) believed this. This result is interesting, because one alleged justification for the Baby Doe regulations is that neonatologists undertreat infants because they do not see how well older children do, but this undercuts such claims. More pediatricians than neonatologists responded that neonatologists overtreat infants.

What Does the Survey Show?

We do not know what all neonatologists or pediatricians believed when we did this survey, but only what those who received and returned the survey said. Arguably, those with the strongest feelings responded, and we do not know what they do, but what they say. Still, responses from almost half, 494, of our nation's practicing neonatologists and 730 practicing pediatricians show that many are very critical of the BDR-II. Still, one cannot automatically conclude something is bad because some people believe it is bad. Experts have been wrong before, and professionals predictably prefer self-regulation and resent government intervention. Moreover, those favoring the regulations might argue that the neonatologists' and other pediatricians' responses to this survey are not evidence that the BDR-II are a mistake, but proof that they are needed: It shows they do not have the proper beliefs or values. However, critical marks from such a large number of neonatologists create at least a presumption that the rules are misguided. This is because, first, neonatologists and other pediatricians are experts in infants' and children's care, and they are dedicated to providing the best possible care for patients (Hippocratic Oath, 1982; Am. Med. Ass'n., 1980). Second, it is reasonable to seek the views of those who must apply policy when evaluating it.

Upon reflection, it is not surprising a larger percentage (49%) of neonatologists than pediatricians (25%) responded. They probably have more familiarity with the issues, cases, and policies. In itself, the low response rate from pediatricians casts doubt on the representativeness of the pediatricians' responses, but it is important to note that 730 pediatricians agreed with half of our nation's neonatologists about the federal regulations on virtually every point. These findings undercut charges that the neonatologists' criticisms of the Baby Doe regulations on our survey (Freeman, 1988) were because of the fact that they did not like regulation of their nursery practices, or that they were overly pessimistic about the prognoses of sick infants because they did not deal with older children.

The Neonatologists, Pediatricians, and the Supreme Court

In what follows, we suggest that there is more than a presumption that the neonatologists and other pediatricians who responded to our survey were correct in criticizing the BDR-II, because their reasons were analogous to those of the Supreme Court in rejecting the BDR-I (*Bowen v. Am. Hosp. Ass'n.*, 1986). The parallel between the BDR-I and the BDR-II is not perfect, however. Differences between them include their bases in law and how they impose duties. Some terms, moreover, may have somewhat different meanings in law and medicine. In addition, the BDR-I have been tested and rejected by the courts, whereas the BDR-II have not. Thus, there is uncertainty about how the courts will view the BDR-II. Nonetheless, the neonatologists' and pediatricians' reasons for rejecting the BDR-II are enough like the Supreme Court's reasons for rejecting the BDR-I to question the propriety of such rules. These reasons are reviewed.

The BDR Are Not Needed

None would deny that there have been cases of medical neglect of infants with life-threatening conditions. The issue is whether new regulations were needed. The Supreme Court concluded that no evidence or reasoning had been offered to support the government's claim that additional regulations were needed to protect handicapped newborns from discrimination: "It has pointed to no evidence that such discrimination occurs" (*Bowen v. Am. Hosp. Ass'n.*, 1986). The Court concluded that even the cases used by the government, "which supposedly provide the strongest support for federal intervention, fail to disclose any discrimination against handicapped newborns" (*Bowen v. Am. Hosp. Ass'n.*, 1986).

Furthermore, the Court writes, by the Administration's admission "*none* of the 49 cases [resulting from 'hot-line' calls to DHHS] had resulted in a finding of discriminatory withholding of medical care!" (*Bowen v. Am. Hosp. Ass'n.*, 1986). The Court further decided that no evidence had been presented to indicate that hospitals or physicians failed to contact authorities when parental choices did not support the child's welfare. In fact, it wrote, the cases presented indicated that they had done so appropriately. Thus, the Court openly questioned the need for such regulations.

Our respondents agreed with the Supreme Court. In the survey, over three-quarters (76%) of the neonatologists and two-thirds (67%) of the nonneonatologist pediatricians stated that the BDR-II were not needed.

Further support for this view comes from a survey from the Office of Inspector General (OIG). It finds ". . . no significant increase in the volume of reports received after October 1985 [when the BDR-II became effective], and most states feel that existing child abuse and neglect procedures would have been adequate to respond to Baby Doe reports" (OIG, 1987).

The BDR Ignore
the Traditional Role of Parental Consent

The Supreme Court held that the administration "does not distinguish between medical care for which parental consent has been obtained and that for which it has not" (*Bowen v. Am. Hosp. Ass'n.*, 1986). When there is a choice between treatments or disagreement among experts about what treatments are best, the parents ought to be able to decide what is best for their child. The Court argued that, "In broad outline, state law vests decisional responsibility in the parents, in the first instance, subject to review in exceptional cases by the state acting as *parens patriae*" (*Bowen v. Am. Hosp. Ass'n.*, 1986). It concluded that no facts had been presented to justify changing this policy.

Neonatologists, in responding to three case histories, overwhelmingly agreed (77–85%) that parental views needed to be considered in setting treatment plans for infants with extremely poor prognosis for survival and cognitive development. Pediatricians who were not neonatologists also believed that parents' views should be considered (73–79%). Yet two-thirds of these neonatologists (66%) and 65% of the nonneonatologist pediatricians believed the current BDR-II affected parental rights to consent to or refuse treatment based upon what was in the infant's best interest.

The BDR Could Cause More Harm than Good by the Poor Use of Resources

The Supreme Court held administrative policy "effectively makes medical neglect of handicapped newborns a state investigatory priority, possibly forcing state agencies to shift scarce resources away from other enforcement activities—perhaps even from programs designed to protect handicapped children outside hospitals" (Bowen v. Am. Hosp. Ass'n., 1986).

Neonatologists and other pediatricians also worried about the proper use of limited funds. Although half of the neonatologists (51%) thought the BDR-II had not exacerbated NICU bed shortages, many were uncertain about this (32%); 17% believed it had. This may reflect regional differences. In contrast, half (50%) of the nonneonatologist pediatricians were uncertain about whether the Baby Doe regulations caused NICU bed shortages; 24% thought it had; 25% thought that it had not. This range of responses may reflect regional differences or the fact that many pediatricians were unaware of the availability of bed space nationally or even in their region. To adopt a nationwide policy requiring treatment without a notion of availability of bed space, however, can create difficult problems about what to do with those outside the hospital walls when all the beds are taken.

An important finding for the Baby Doe controversy was that most neonatologists and pediatricians responded that severely sick infants with a poor prognosis were overtreated in NICUs. Even more nonneonatologist pediatricians (62%) thought this was true than neonatologists (56%). This undercuts claims that neonatologists are not as aggressive as they should be in treating very sick infants because they are unfamiliar with their outcomes as older children.

The high cost of intensive care coupled with limited funds for children's health-care programs may have led over four-fifths of the neonatologists to respond that, if life-saving treatments were required for infants with such bleak prognoses, the government should pay for their acute care (82%) and for their subsequent rehabilitative care (82%). Three-quarters of the pediatricians who were not neonatologists agreed, responding affirmatively to both items (75–76%). Some commented that the DHHS grants to the states through the BDR-II did not begin to cover these costs.

The BDR Exert Undue Pressure

The Supreme Court ruled that unwilling state agencies might be under undue pressure to comply with the BDR-I. Even an impartial state agency could "still be denied federal funds for failing to carry out the Secretary's mission with zeal" (*Bowen v. Am. Hosp. Ass'n.*, 1986). The Court also said:

> For while the Secretary can require state agencies to document their *own* compliance with Section 504, nothing in that provision authorizes him to commandeer state agencies to enforce compliance by *other* recipients of federal funds (in this instance, hospitals). State child protective services agencies are not field offices of the DHHS, and they may not be conscripted against their will as the foot soldiers in a federal crusade (*Bowen v. Am. Hosp. Ass'n.*, 1986).

Although the legal basis of the current BDR-II is different, arguably they also exert pressure on state agencies. First, although states may refuse the federal funds, and even agree it might be cost-effective to do so, some high state officials privately express fear they will lose favor with DHHS if they turn down these funds. Other funding requests could be jeopardized, they worry, if they are seen as uncooperative with what even the Supreme Court calls "a federal crusade" (*Bowen v. Am. Hosp. Ass'n.*, 1986). Second, they say they have greater needs for funds elsewhere. The OIG survey reflects this view, reporting that many states believe that they "should have the flexibility to use available funds for medical neglect in general or other priorities if needed" (Steinhaus, 1986).

The administration, however, responded that there was no intention to create undue pressure. They state that neither the BDR-I or the BDR-II were intended to alter standards of care, require treatments, or undermine parental rights. For example, the Solicitor General argued that the government was "essentially concerned only with discrimination in the *relative* treatment of handicapped and non-handicapped persons and [this] does not confer any *absolute* right to receive particular services or benefits under federally assisted programs" (*Bowen v. Am. Hosp. Ass'n.*, 1986). The Supreme Court, however, disagreed stating, "The Final Rules imposed just this sort of absolute obligation on state agencies that the Secretary had previously disavowed ... they constitute an *absolute* right to receive particular services or benefits under a federally assisted program" (*Bowen v. Am. Hosp. Ass'n.*, 1986).

Neonatologists also indicated that the BDR-II exerted undue pressure, albeit of a different kind. In responding to the cases, 22–33% said that they practiced differently as a result of the BDR-II. In each of the three cases, there were significant differences between responses to what neonatologists viewed as the BDR-II requirements and as proper treatments. In Case 3, 32% of the respondents said that life-saving treatment was not in the infant's best interest, but that the BDR-II required it. Most (55%) also said that infants with very poor chances for survival were being overtreated in NICUs. This suggests that overtreatment of such infants may be a problem resulting from the BDR-II.

Pediatricians who were not neonatologists also indicated that the BDR-II exerted undue pressure of the sort that affects practice decisions. In reporting on the cases, 36–38% said that they practiced differently as a result of the BDR-II. In two of the three cases (Cases 1 and 2), there were significant differences between responses to what pediatricians viewed as the BDR-II requirements and as proper treatments as well. Even more pediatricians (62%) than neonatologists said that infants with very poor chances of survival were being overtreated in NICUs. Thus, both groups seem to believe that overtreatment of such infants may have been a problem because of the BDR-II. If the BDR-II alter court rulings (Order in the Steinhaus Case, 1986) or neonatal practice, this will affect all neonatal care standards even in those states that reject Baby Doe funds.

The BDR Criteria Are Problematic

The Supreme Court in *Bowen v. Am. Hosp. Assoc.* rejected the administration's views that it is prejudicial and unjust to evaluate the infant's handicap in selecting a treatment plan. Quoting the Appeals Court in the Baby Jane Doe case, the Court argued that "where medical treatment is an issue, it is typically the handicap itself that gives rise to, or at least contributes to, a medical treatment decision" (*Bowen v. Am. Hosp. Ass'n.*, 1986). If children who were "otherwise qualified" were not treated because of their race, the Court said, this would violate Section 504, but those infants with multiple handicaps or extreme prematurity may not, precisely because of those conditions, be otherwise qualified for treatment.

Even with the relevant portion of the BDR-II before them, neonatologists and other pediatricians disagreed about what the BDR-II required. In three hypothetical cases of infants with profound cognitive damage, a life-threatening condition, and a very poor chance of survival, there was little agreement about what was required. Some neonatologists (14–18%) were uncertain, whereas some held that treatment was required (21–46%), and some thought that it was not (38–60%). There was even more uncertainty among the other pediatricians. Some (18–27%) were uncertain, whereas some held that treatment was required (40–53%), and some thought that it was not (27–32%).

It is not surprising that there is such a lack of agreement as to what is required by the Baby Doe regulations, because the exceptions given in the rules are problematic. The *first* exception is that the infant "is chronically and irreversibly comatose." There is, however, great uncertainty now about what medical criteria could be used to determine this in the newborn period (Coulter, 1987). Further confusion arises because a trial judge finally ruled that a baby who was in a persistent vegetative state met this condition, even though the pediatric neurologists denied he was chronically and irreversibly comatose (Gianelli, 1986). Determining which of several states of permanent unconsciousness the infant suffers from may be a difficult, technical decision, and it is not clear why distinguishing among them is so relevant in deciding whether nontreatment is justified.

The *second* exception is that treatment would "merely prolong dying, not be effective in ameliorating or correcting all of the infant's life-threatening conditions, or be otherwise futile in terms of the survival of the infant." This is problematic because first, it is difficult to determine what "prolonging dying" means especially in an NICU; neonatologists could in many cases, with a maximal technological effort, forestall death for a very long time. Second, a life prolonged in this way might be so filled with suffering that none would wish it for anyone. To treat others as you would wish to be treated is one well-established test of a moral act. Thus, it may be that the BDR-II sometimes require neonatologists to do unto others (prolong suffering for the sake of biological survival alone) what they would not do for themselves or their children. Third, there is confusion about how to view such treatments. In medical science, to claim knowledge usually requires a large sample of patients and a high confidence

level. For some of the most heroic treatments on the sickli-
est patients, however, this kind of proof is often lacking. Some
may suppose that, without this proof of futility, the BDR-II
require them to provide heroic therapeutic care. Others may
supopose that, without this proof, they are engaging in inno-
vative or experimental therapy that requires special research
review and consent. Fourth, it is not clear why prolonging
dying is, in itself, such an exceptional consideration. In some
cases, when there is not great suffering those extra days,
weeks, or months are precious.

The *third* exception is that "the provision of such treatment would be virtually futile in terms of the survival of the
infant and the treatment itself under such circumstances
would be inhumane." It was said to be a compromise condition, because it seemed to permit some consideration of
suffering and the nature of the infant's life. Strictly speaking, however, this is not so, because it allows consideration
of virtual futility and inhumanity only in terms of biological
survival. Seen in these terms, there is little to distinguish
the restrictions presented in the BDR-I and the BDR-II. Moreover, the DHHS Appendix suggests that this condition must
be interpreted narrowly; that is, it is an allowable consideration only when the infant's death is imminent (DHHS, 1985).
The Appendix does not have the same force as the regulations themselves, but it contributes to an uncertainty about
how much latitude physicians and families are permitted
after they determine treatment is futile and inhumane, but
before they are convinced death is imminent.

Another source of confusion is the relation of the three
exceptions to the statement that the BDR-II forbid ever discontinuing "appropriate hydration, nutrition and medication."
This leaves it uncertain as to how to treat an infant with a
life-threatening condition *even though* he or she meets one
or more of the exceptions. It would be a stark contrast to standard medical care, for example, to treat an anencephalic
infant with antibiotics for life-threatening meningitis. Is
it required by the BDR-II or not? One might say it *is not*

required, because he or she has a life-threatening condition that cannot be corrected, but one might also argue it *is* required, because the BDR-II say that appropriate medication must be given even if one of the conditions has been met. One problem is that "appropriate" is ambiguous. It may mean that it is the *appropriate treatment for a kind of condition.* (It is appropriate to give antibiotics for meningitis.) In contrast, it may mean that it is the *appropriate treatment for a particular patient with that condition.* (It is not appropriate to give antibiotics to an anencephalic infant with meningitis.) Traditional medical judgment looks at the utility or benefit to a particular patient of the proposed treatment.

In general, these objections raise variations of the same problems. First, relevant considerations are ignored by these regulations. It seems reasonable to consider not only biological survival alone, but also prognosis, pain and suffering, treating others as each of us wants to be treated, harms and benefits, and parental views in making treatment choices. Second, these rules seem intrinsically problematic.

The BDR Seek to Change the Standards of Care

The Supreme Court (*Bowen v. Am. Hosp. Ass'n.*, 1986) and the Second Court of Appeals of the Second Circuit (University Hosp., SUNY, 1984) have concluded that the purpose of the BDR-I was, despite disclaimers, to alter standards of care. They were critical of such attempts. Judge Leonard D. Wexler wrote, "the government has taken an oversimplified view of medical decision-making" (University Hosp., SUNY, 1984). Judge Gerhard Gessel stated that the regulations were "arbitrary and capricious" (*AAP v. Heckler*, 1983) and "intended . . . to change the course of medical decision-making" (AAP vs. Heckler, 1983). He also said that they sought to require physicians to give maximal treatment for every nondying infant without considering parental views or the suffering it might cause.

The BDR-II are also widely seen as seeking to alter standards of care. Defenders might argue that physicians are required only to use reasonable medical judgments: those "that would be made by a reasonably prudent physician, knowledgeable about the case and the treatment possibilities with respect to the medical conditions involved" (DHHS, 1985). Furthermore, in the Appendix, DHHS seeks to rebut charges that the BDR-II will cause overtreatment or overvaluation of infants on the grounds that it defers to reasonable medical judgment. It defers, however, only to the fact that physicians, ethics committees, or Child Protective Service (CPS) investigators are needed to make the *technical* decisions about whether the conditions have been fulfilled. In contrast, reasonable medical judgments are *traditionally* seen in broader terms as including *moral as well as technical* considerations about what is harmful or beneficial to patients, and what ought to be done to take care of someone in accordance with that person's best interests.

The survey, moreover, shows that less than one year after the BDR-II became final, the regulations apparently have altered care practices. In responding to one case, a third of all pediatricians, 32% of the neonatologists, and 36% of the others said that they have changed how they practice medicine as a result of the BDR-II. In comparing responses to two items, we find that up to a third regard the BDR-II as requiring them to act in a way different from what they consider the infant's best interest.

The BDR-II Undermine
the Best-Interests Standard

There is a long-standing, well-accepted medical and legal tradition to use a best-interests standard for determining how physicians and courts should respond to children's needs (Holder, 1983). The Supreme Court in *Bowen v. Am. Hosp. Ass'n*, for example, stated that the Baby Jane Doe case failed to show any need for regulations, because the best-interests standard was followed. It agreed with ". . . the

Appellate Division which found that the concerned and loving parents had chosen one course of appropriate medical treatment over another and made an informed decision that was in the best interest of the infant . . ." (*Bowen v. Am. Hosp. Ass'n.*, 1986). Thus, the Supreme Court applied the best-interests standard.

Most neonatologists responding to three cases of critically ill and profoundly cognitively impaired infants with a poor prognosis for survival overwhelmingly agreed (75–86%) about what was in the child's best interest, but in those same cases, up to a third perceived a conflict between their duty to act in the infant's best interest and the legal requirement to act in accordance with the BDR-II. In each case, too, there was a significant difference (p = <0.001) in their responses to what was the best way to take care of these infants and what the BDR-II required. Over half (56%) believed very sick infants were being overtreated in NICUs. The pediatricians' responses were similar with the exception of Case 3.

In response, one could argue that these regulations do not require physicians to inflict nonbeneficial "treatments." The BDR-II are optional funding regulations that impose a duty on state agencies to set certain laws, procedures, or policies to monitor possible medical neglect. Still, we have seen that many neonatologists see the BDR-II undermining the best-interests standard. In their comments, many worried about not only infants' overtreatment, but also the families' suffering.

We share the concern that the BDR-II can cause unnecessary pain to infants and their families. Suppose that a family with the advice of physicians decides that it is in the best interest of their severely sick and profoundly cognitively damaged infant with a poor chance of survival to forgo heroic therapy. Suppose too that the infant's condition fails exactly to meet the exceptions stated in the BDR-II. Some hold that the approval of a hospital ethics committee would allow withholding of treatment, but unless state law explicitly allows this, committee members are no less obligated to follow the BDR-II than other citizens. Another possibility is that every departure from BDR-II must go to the courts for review. Others hold that the proper procedure in such cases is to report the case to the state's CPS. It is their duty to investigate and determine neglect. Suppose they do not find neglect. A similar dilemma then arises for CPS, since they are instructed to follow the BDR-II. Should they follow the best-interests standard or the Baby Doe standard, when they are different, or should they go to the courts? In any case, the family's grief would be confounded by the fact that they were reported and investigated for possible medical neglect of their infant. This is a source of suffering and loss of confidentiality for the family that seems unnecessary without *evidence* that there is a genuine need to monitor parental choices, attending physicians, or ethics committees.

The potential for conflict between the BDR-II and the best-interests standard exists. For example, Baby Lance was in a persistent vegetative state. He was an infant who had been beaten by his father. His mother, physicians, and the Ethics Committee at the University of Minnesota Hospital decided that he should be given only comfort care. Lance's physician testified that heroic or life-prolonging therapy would not be in his best interest. Yet the judge ruled ventilator support and antibiotic therapy had to be provided, since the BDR-II exceptions had not been met. He later ruled Baby Lance could be given only comfort care because he was chronically and irreversibly comatose (Order in the Steinhaus Case, 1986). The pediatric neurologists, however, denied that Lance was chronically and irreversibly comatose (Gianelli, 1986). We do not know how higher courts would view this ruling, or assess this potential conflict between the best-interests standard and the BDR-II.

The survey shows that many pediatricians, and neonatologists believe that the BDR-II do not serve the best interests of infants. A basic moral tenent of medical practice is that one should act in the patient's best interest, so far as this is possible. As many as 33% of pediatricians, however, say that they have altered their standard of care as a result of the new federal regulations. They say the BDR-II can require heroic treatments that are not in the patient's best interest. A majority (56%) of neonatologists responded that infants were overtreated; even more pediatricians (62%) said infants with poor prognoses for survival were overtreated in our NICUs. This is a significant finding for the Baby Doe debate. If true, it shows that neonatologists do not undertreat infants because they are unfamiliar with how older children thrive once outside of their NICUs. We saw that many were concerned about the use of resources and possible bed shortages. If these regulations promote or require special, heroic, or burdensome treatment for some of our citizens but not others, then these regulations might be unconstitutional because they fail to protect us all equally. More than half of both pediatricians in nonneonatal fields and neonatologists state that the BDR-II offer inadequate consideration of parental views; a majority of both groups also think it allows too little consideration of the infants' suffering. We do not believe these consequences were the intent of Congress in passing the 1984 Child Abuse and Treatment Amendments.

Acknowledgments

This chapter contains additional material about pediatricians, but is adapted from information appearing in the *New England Journal of Medicine* by Loretta Kopelman, Thomas Irons, and Arthur E. Kopelman: Neonatologists judge the "Baby Doe" regulations. (1988) **318,** 677–683. We want to thank Ross Laboratories for supporting this project, and Michael D. Cruze and Donald Holbert for help with the statistical analysis of the data.

References

American Academy of Pediatrics (AAP) v. Heckler, 561 F. Supp. 395, 397 (D.D.C.) 1983.
American Medical Association (1980) Principles of medical ethics. *Am. Med. News* **66,** 9.
Avery, G. B., ed. (1987) "Neonatology: Perspective in the Mid-1980's," in *Neonatology: Pathophysiology and Management of the Newborn,* 3d Ed. (J. B. Lippincott, Philadelphia), pp. 9–12.

Bowen vs. American Hospital Association, et al. (1986) US Supreme Court. 106 S.Ct. 2101, No. 84-15-9.

Coulter, D. L. (1987) Neurological uncertainty in newborn intensive care. *N. Engl. J. Med.* **316,** 840–844.

Department of Health and Human Services (DHHS) (1984) *Non-discrimination on the Basis of Handicap Relating to Health Care for Handicapped Infants*. 49 Fed. Reg. 1622-54 (Referred to as BDR-I).

Department of Health and Human Services (DHHS) (1985) *The Child Abuse and Treatment Amendments of 1984* (Public Law 98-457). 50 Fed. Reg. 14878-901 (Referred to as BDR-II).

Freeman, J. M. (1988) "Baby Doe" Regulations. *N. Engle. J. Med.* **319,** 726.

Gianelli, D. M. (1986) Minnesota judge: Baby Lance can die. *Am. Med. News* **1,** 34.

The Hippocratic Oath, Ancient (1982), in *Contemporary Issues in Bioethics*, 2d Ed. (Beauchamp, T. L. and Walters, L., eds.), Wadsworth, Belmont, CA, p. 121.

Holder, A. R. (1983) Parents, courts and refusal of treatment. *J. Pediatrics* **103,** 515–521.

Kopelman, L. M., Irons, T. G., and Kopelman, A. E. (1988) Neonatologists judge the "Baby Doe" regulations. *N. Engl. J. Med.* **318,** 677–683.

Office of Inspector General (OIG), Office of Analysis and Inspections (1987) *Survey of State Baby Doe Programs* (Publication no. OA1-03-87-0018).

Order in the Steinhaus Case (1986) *Issues Law Med.* **2,** 241–251.

Smith, D. W. (1982) Recognizable patterns of human malformations: Genetic, embryologic and clinical aspects, in *Major Problems in Clinical Pediatrics*, 3rd Ed., vol. 7. (W. B. Saunders, Philadelphia).

University Hospital, State of New York at Stoney Brook (SUNY), US Court of Appeals of the Second Circuit, No. 679, Docket 83-6343 (1984).

US Child Abuse Protection and Treatment Amendments of 1984, Public Law 98-457. Amendments to *Child Abuse Prevention and Treatment Act* and *Child Abuse Prevention and Adoption Reform Act*. Amendment No. 3385. Congressional Record 1984; Senate S8951–S8956.

US Rehabilitation Act. Public Law 93-112. 29 USC. 794.

Zar, J. H. (1984) "Testing for Goodness of Fit," in *Biostatistical Analysis*, 2d Ed. (Prentice Hall, Engelwood Cliffs, NJ), p. 45.

The Impact
of the Child Abuse
Amendments on Nursing
Staff and Their Care
of Handicapped Newborns

Joy Hinson Penticuff

This chapter will discuss:

1. The impact of misinterpretation of the Child Abuse Amendments on infant treatment decisions, particularly in reference to "instrumentalists' " overtreatment of extremely premature infants;
2. How nurses' and physicians' interactions with infants may result in differing conclusions about the success or futility of treatment, thus producing different opinions about the relevance of the amendments;
3. The impact of overtreatment resulting from the amendments on nursing staff morale, especially in terms of erosion of a sense of ethical integrity; and
4. The impact of overtreatment on nurses' relationships with physicians and with families.

From: *Compelled Compassion* Eds.: Caplan, Blank, and Merrick
©1992 The Humana Press Inc.

The Child Abuse Amendments have had two easily detectable effects on neonatal intensive care and one effect that is more subtle, but from a nursing perspective, far more significant. The obvious effects are, first, nurses' increased awareness that life-sustaining medical treatment cannot be withheld unless such a course is sanctioned by hospital authorities and, second, the establishment of Infant Care Review Committees (ICRCs), which usually function to provide or withhold such sanction. The more subtle impact of the Child Abuse Amendments has been their insidious reinforcement of a sense that, in the Neonatal Intensive Care Unit (NICU), we are obligated to unleash our full technical armamentarium to preserve life. It is this last, implicit, aspect of the amendments that I wish to examine, because it threatens to overwhelm the humanistic perspective held by many NICU nurses.

This chapter argues that the amendments have fostered "instrumentalist" (Breshnahan, 1987) attitudes and have impeded the emergence of "humanistic" attitudes in neonatal intensive care. In extreme form, the instrumentalist attitude is that life must be saved at all costs and in all circumstances. Against this, the humanistic attitude, as characterized by McCormick (1974), is that life is not an end in itself, but a vehicle through which one realizes other, more transcendent ends, such as relationships.

Butler (1986, 1989) offers an insightful description of the "instrumentalist" view:

> The infant is primarily "a body in which there are difficult technical problems to be solved", and the focus is on full technologic support of body physiology. "The only real power is the power to save lives, and it seems unreasonable purposely to allow death. . . . No one wants a baby to die on his or her shift—even if prolonging life means prolonging dying. . . . In this milieu, it is inappropriate to ask the question, What makes life valuable?" (Butler, 1986).

Many nurses reject the instrumentalist view as too narrow in its portrayal of the infant and the true intent of neonatal care. A view more congruent with nursing's traditional humanistic values is that the infant is a newborn human being (not primarily a body), with the capacity to experience comfort and pain, with a social role (newborn son, daughter), and with a fragile developmental potential that must be protected if the infant is to flourish. In this humanistic view, we fail to meet our ethical obligations to the infant if we do not ask the question: What makes life valuable?

The reality of neonatal intensive care is that humanistic approaches are easily crushed by application of a technology that can measure the physiology of a beating heart, but cannot measure the courage and compassion that symbolically are held within that heart. Granted, the instrumentalist attitude has not held sway solely because of the amendments; instrumentalism was inherent in the early creation of NICUs as places in which high technology and scientific training of staff combined to save infant lives. However, the language of the amendments, with the focus on "medical indications," is taken by instrumentalists to mean that, if a treatment holds even a remote possibility of saving life, it must, under law, be used. The amendments allow a justification for what can, for some infants, be little more than prolonged torture. Weir (1984) describes such a case:

> Mignon was an extremely premature infant. Born after only 19 weeks' gestation, she weighed 482 grams at birth [about one pound]. Nevertheless, she was admitted to an NICU for treatment. She encountered multiple problems brought about by her prematurity and low birthweight, but continued to live with aggressive medical treatment. After seven months in the NICU, she was described by her physician in the following manner: she has been plagued with frequent kidney failures, liver problems that defy textbook descriptions, and the usual lung problems. . . . at seven

months after delivery she is hanging in there. We have
kept the [umbilical artery] catheter in her aorta all
the way, so her hyperalimentation [intravenous
nutrition] has been intra-arterial. . . . she can't suck
because of the endotracheal tube [tube from venti-
lator]. . . . of course, she has been constantly on PEEP
[positive pressure ventilation]. . . . But with no spontaneous breathing, all mechanical ventilation, her
lungs are pretty damaged now. . . ." (explanatory terms
in brackets).

As time has passed since their implementation, the
specific content of the amendments has faded from the memo-
ries of NICU nurses, but remaining is a strong impression
that it may be illegal to withhold any life-sustaining therapies from any infant. Today it is rare for NICU in-service
training to include the amendments as part of orientation to
the care of critically ill newborns (Penticuff, 1990). In contrast, immediately after promulgation, some discussion of the
amendments—frequently unclear and misleading—was usually included in the in-service education of nursing staff.
Nurses who have been in neonatal intensive care units for
several years remember the furor caused by the initial regulations, which required the posting of a sign in each NICU
stating that discriminatory failure to provide medically indicated treatment was against the law. Novice nurses are less
aware of the amendments, but most NICU nurses—novice
and experienced—have been exposed to mass media regarding handicapped infants. Also, most recognize that failure to
support the life of a handicapped newborn, without proper
review by appropriate institutional officials, is illegal.
  A major fly in the ointment is that the amendments do
little to help nurses deal with the most frequently encoun-
tered NICU dilemma: whether and how aggressively to treat
the borderline-viable extremely premature infant of less than
27 weeks' gestation, the infant like Mignon. Generally speaking, how such an infant's treatment is managed depends most
strongly on whether the attending neonatologist is an instrumentalist or a humanist. The secondary influence is likely to

be the residual understanding of the physicians about what hospital attorneys have interpreted as medical treatments required by the amendments. The third influence is the strength of the nursing staff's consensus about the infant's response (positive or negative) to the medical treatment.

Of course, instrumentalist and humanistic approaches vary from one NICU to the next, and unfortunately, the interpretations given the Child Abuse Amendments also vary because the language of the amendments practically guarantees that differing interpretations can be drawn. The next section deals with the impact on nursing staff of misinterpretation of the amendments.

Misinterpretation of the Language of the Amendments

The portion of the Child Abuse Amendments that produces the most confusion is the language that allows not providing life-sustaining medical treatment when "the provision of such treatment would be virtually futile in terms of the survival of the infant, and the treatment itself under such circumstances would be inhumane" (Child Abuse Amendments, 1984). When are therapies "virtually futile"? The instrumentalist and the humanist are likely to disagree on what counts as virtually futile, and because intensive care for extremely premature infants usually entails invasive, iatrogenic therapies with uncertain outcomes, the "inhumaneness" of therapy is also a point of contention. The instrumentalist and humanist are likely to disagree on how much burden is acceptable, given the uncertainty of outcomes.

The more humanistic medical and nursing staff may see the amendments' language as meaning that an extremely premature infant, born at 25 weeks' gestation, who has suffered grade four (serious) intracranial bleeding and who is still requiring mechanical ventilation on 100% oxygen and high ventilator settings at six months of age, can be withdrawn from life-sustaining therapy because such therapy is "futile and inhumane." They see life-sustaining therapy as merely prolonging the dying process for this infant. Instrumentalists may interpret the same language as mandating continued life-sustaining treatment for the same infant, because they believe that the treatment is not futile and they do not regard it as inhumane.

Nurses and many of the pediatric residents who rotate through NICUs usually do not understand the amendments except as mandating treatment. They have not read the actual amendments and usually have not had the benefit of discussions with someone who has studied them in depth. This often results in application of aggressive therapies that can be seen as constituting "overtreatment." If the neonatologist is an instrumentalist, the amendments provide a convenient reason to try even the most improbable treatments to sustain the life of a struggling borderline-viable premature infant. For the more humanistic neonatologist, the amendments present the threat that government regulations may necessitate the provision of treatment that will save the biologic life of the infant, but cannot save the essential developmental potential of the child.

The fault of the Child Abuse Amendments is their disregard for quality-of-life issues. If we take quality of life from the infant's perspective, we can surely conclude that a life of pain without compensating affectionate comforting by someone who loves the child is not a good for the child. Adults have the ability to refuse medical therapies that would be burdensome or would interfere with their attainment of life goals, but parents of infants in NICUs are rendered essentially powerless under the amendments in regard to consent to medical therapies for their infant. Parents who clearly love their baby, who understand the infant's condition and prognosis, and who, after consultation with medical experts and family advisers, decide that they do not wish to consent to therapies that will sustain a life not beneficial to their child are stripped of decision-making prerogative. They may have access to an ICRC, yet these committees vary widely in their functioning (Fleming et al., 1990; Fleischman, 1986). There is no system of checks and balances. In some hospitals, the committees are very willing to grant the reasonable requests of families who obviously have the infant's best interest at heart. Other committees are more likely to be influenced by slim probabilities of survival, and are reluctant to consider family values and "unscientific" views of what is in the infant's best interest. Some committee decisions are vetoed by hospital attorneys who fear the threat of the amendments even though the ICRC has sanctioned parental requests for withdrawal of life-sustaining treatment.

It is ironic that the amendments, which supposedly protect infants from child abuse, may in some cases actually be the instrument for child abuse in the form of prolonging dying under conditions of a brutal intensive care from which there is no escape. Most experienced nurses believe that intensive care is inherently damaging for infants, in that there is infliction of pain without the ability to comfort the baby adequately. There is separation of the infant from the normal experiences of parental caretaking, and this may impede the usual process of parent–infant attachment. There is both deprivation of the usual tender, need-fulfilling "mothering" of the infant during the NICU stay, and the experience of overstimulation of noxious or painful invasive stimuli—emergency needle thoracentesis, uncountable heel sticks, numerous insertions of needles to start intravenous lines, prolonged periods of sleep interruption, periods of anxiety caused by acute air hunger, and other discomforts. The sicker the infant, the longer the NICU stay, the more damaging the physical and emotional cumulative effect of these experiences (Greenspan and Porges, 1984; Gorski, 1984; Gorski et al., 1980; Gottfried and Gaiter, 1985).

Is Therapy Futile?
Nurses and Physicians May Disagree

I now turn to another issue relevant to the interpretation of some of the vague language of the amendments. Nurses may view an infant's response to therapy as indicative that therapy is futile, whereas neonatologists may view the infant's response as indicative that therapy is possibly of benefit. How can this difference come about? The key is in understanding how nurses' and physicians' interactions with infants differ. Nurses care for the infant over eight- to twelve-hour shifts, and their care occurs within arm's length of the infant. Thus, nurses have the most continual and close exposure to infants' responses to treatment of any member of the health-care team. Nurses are trained to make highly discriminant observations about the infant; indeed, nursing expertise in neonatal assessment is a requirement without which neonatal intensive care cannot be carried out.

In an important study by Anspach (1987), nurses were found to use additional criteria to those used by neonatologists in drawing conclusions about infant prognosis. Nurses attend to the same technical criteria used by physicians: physical appearance, laboratory data, ventilator settings, and intensity of care required, but the nurses use, in addition, the infant's subtle behavioral cues, such as levels of alertness, visual regard or gaze aversion, and responses during interaction, as additional data on which to base prognosis. Anspach concluded that nurses may be overly sensitive to these subtle infant cues, and physicians may be overly insensitive to them. The point here, though, is that this attention to different criteria may result in differing evaluations of treatment response and prognosis, thus producing different opinions of the relevance of the Child Abuse Amendments. For the nurse, the language of the amendments excluding therapies that are virtually futile and inhumane may be seen to apply to the infant, thereby permitting withholding of life-sustaining therapy. However, the physician, experiencing the same case differently, may hold a totally opposing view.

A study of perceptions of ethical problems by nurses and doctors by Gramelspacher et al. (1986) corroborates Anspach's findings. Gramelspacher et al. found important differences about how aggressively medical therapy should be pursued, often including decisions about whether or not to resuscitate a patient and how vigorously to give pain medication. They conclude:

> Potential conflicts are exacerbated because nurses' close contact with patients leads them to see the results of medical intervention far more intensely than do physicians. Also, physicians see themselves as accountable to other physicians, and to patients and their families, but not to nurses. They may not recognize conflicts with nurses, and therefore not attempt to resolve these conflicts. This failure to address differences of opinion in the intense work setting of a hospital ward impedes optimum, humane medical care (Gramelspacher et al., 1986).

Another aspect of the nurses' care of the infant also significantly influences nurses' views about whether treatment is inhumane: nurses' closeness to the inflicting of pain. It is nurses who most frequently carry out the routine painful, invasive procedures—starting intravenous, sticking heels for blood sampling, inadvertently peeling off the skin when removing monitor leads—required by current neonatal technology. In other instances, when nurses do not carry out the procedure, they assist the physician. In assisting, it is the nurse's role to observe the infant's responses to the procedure (respiration, heart rate, color change, or hemorrhage, for example) while the physician focuses on insertion or manipulation of instruments. Procedures performed day in and day out in a busy NICU include emergency chest tube insertions to reinflate collapsed lungs—usually done without benefit of anesthesia, analgesia, or sedation—needle puncture of arteries lying close to the skin for blood sampling—again, usually without benefit of pain relief (Franck, 1987), and countless episodes of endotracheal tube suctioning, which clears the infant's airway, but induces significant momentary discomfort.

Until the late 1980s, it had been a widespread practice for infant surgery to be performed without benefit of anesthesia (Berry, 1986; Butler, 1987). Paralysis was achieved through use of pancuronium bromide (Pavulon), but no pain relief whatsoever was provided for major or minor surgeries. This practice, decried by one neonatologist as "barbarism," is now beginning to wane because of parent and nurse protest (Scanlon, 1986; Lawson, 1986,1988). It should come as no surprise that nurses inevitably begin to question the morality of burdensome treatment if there is not some minimal hope for corresponding benefit to the infant undergoing neonatal intensive care.

Impact of Perceived Overtreatment on Staff Nurse Morale

I have argued in the preceding sections of this chapter that the Child Abuse Amendments have played a subtle, yet powerful, role in maintaining an instrumentalist attitude in many NICUs in this country. The instrumentalist attitude was necessary to the advances of neonatal medicine and the rescue of thousands of infant lives. When neonatal intensive care was first begun, life was assumed to be good for all babies, and the preservation of life through application of high technology and science was justified because life was assumed to be a benefit for babies. We learned slowly, through developmental followup (Escalona, 1982; Sameroff, 1981; Schraeder, 1986; Schraeder et al., 1987), that life was not necessarily good for all babies who survived the NICU. We learned that the smaller infants rarely escaped neurologic impairment, ranging from minimal, such as learning difficulties, to profound, in which human interaction was beyond the child's capacity (Baerts and Meradji, 1985; Bernbaum et al., 1984; Calame et al., 1985; Kraybill et al., 1984; McCormick, 1985; McMenamin et al., 1984; Rice and Feeg, 1985; Stewart et al., 1983; TeKolste et al., 1985; Walker et al., 1984; Williamson et al., 1983; Yu et al., 1986). We

learned that, for some infants, technology is not able to prolong life without destroying the infant's potential for a good
life. The entirety of life is not lost, only parts of it. The Child
Abuse Amendments mitigate against the difficult analysis—
which some of us see as our obligation—of whether what
remains of life can be good for the infant.

Neonatologists and nurses who reject the notion that life
is good for all babies seek a more humanistic approach to
neonatal care, in which family values and infant quality of
life can be considered in decisions to apply life-sustaining
therapies. Duff (1987) suggests that we should emphasize
the primacy of human relationships, and the importance of
feelings and family values as we weigh our obligation to seek
what is best for infants in intensive care. The Child Abuse
Amendments force us to turn in the opposite direction, with
their specific narrowing of considerations to "medical indications" as we attempt, with families, to decide what is best for
imperiled infants.

The technology of the NICU—ventilators, intravenous
nutrition, and drugs—is highly effective in forestalling death,
although as mentioned before, the iatrogenic complications
(intracranial hemorrhage, blindness, liver and kidney failure, to name just a few) occur as side effects. Thus, a consequence of using the narrow "medical indications" criterion is
overtreatment, because use of the technology can forestall or
prevent death, and an instrumentalist physician will see the
possibility of preventing death as an indication to treat.
Unfortunately, overtreatment has a negative effect on nursing staff morale, especially in terms of erosion of nurses' sense
of ethical integrity (Mitchell, 1983), and ability to carry out
what they see as their ethical obligation to avoid doing harm
to infants and families in the NICU (Penticuff, 1990).

Although nurses are called upon to implement treatment
plans that they may see as constituting burdensome
overtreatment, their input into decision making is usually
not sought and, in some situations, is actively disregarded.
Some physicians respond punitively when questions are
raised about the humanity of treatment that may result in

infant survival, but with profound mental and physical impairment. Nurses experience frustration, anger, and sometimes moral distress and moral outrage (Wilkinson, 1987) when their input is rejected or discounted in medical decisions in the NICU. To be a part of treatment that you believe is torture for the infant—based on your own observations of the infant's continuing pain, agitation, and air hunger—produces both ethical anguish and an erosion of one's personal sense of ethical integrity. Your inner conscience says, "I continue to be a part of something evil." Staff burnout and turnover frequently result (Jacobson, 1978; Cameron, 1986; Penticuff, 1985,1990).

Impact of Overtreatment on Nurses' Relationships with Families and Physicians

When nurses believe over a prolonged period of time that what they are doing is not congruent with what they believe to be right for the infant and family, a profound sense of disillusionment with nursing and medicine often results. This disillusionment can take many forms, but it almost inevitably changes the way nurses interact with those physicians who continue to press for aggressive therapies. More experienced, secure nurses may attempt to challenge the physician or to enlist the support of others to influence the instrumentalist to change the treatment plan. However, many nurses do not have the security to "make waves" within the NICU (Dennis, 1983). Probably the most significant element is the influence and power of the nursing department in the hospital. When staff nurses feel that they have administrative support and protection, they are much more likely to challenge medical treatment that they see as harmful to infants and families than if nursing is not powerful within the hospital (Penticuff, 1990).

When a group of nurses has misgivings about the "good" of therapy for a particular infant, a power dynamic often arises within the NICU. The physician attempts to exert his or her "will" over the nursing staff. Parents may be isolated on the fringes of this dynamic, feeling—but not clearly understanding—what is taking place in the process of decision making about their infant's care. Butler (1986) provides an excellent description of the humanistic and instrumentalist NICUs, contrasting the valuing of parental involvement in decisions and the discounting of their involvement. First, the instrumentalist approach to the parents:

> The parents of these babies are part of the periphery. Obviously, they have no expertise. Although they are expected to be interested in their babies, the mother and father are not to be overly appalled by the myriad of wires, tubes, and tape covering their nevertheless naked, spread eagled child, preventing full body contact of parent to child; by the unnatural noises of beeping and slushing monitors and laughing or tense and anxious voices, always of adult to adult; by the blue, swollen, multiply pricked heels of their baby's feet. . . . Parents must accept the NICU's belief that infants this small really have discomfort rather than pain. Anyway, the baby's tiny organs could not detoxify pain-killing drugs. Good parents are grateful for the level of care available to them, understand, and conform to the unit's practices and schedule. Input from parents about care decisions is suspect because they may be suffering from shock and grief. Moreover, parents are at the bottom of the hierarchy of the health care team (Butler, 1986).

The humanistic NICU, in contrast:

> Parents feel welcome on the unit. They are given complete medical information upon their child's admission and regularly thereafter. What happens when

relief of the suffering of some newborns probably cannot come from curing? The parents, physicians, and
other team members decide together, using the values of all, with the families' values uppermost, what
course of action to take. They take time, allowing the
decision to be a process (Butler, 1986).

The issue of parental involvement in life-and-death
decisions in the NICU is similar to the issue of nurse involvement, in that parents and nurses may have little influence.
The decision-making prerogatives of parents are linked to
the neonatologists' views about how much input parents
should have (Harrison, 1986). Parents typically trust that
physicians and nurses will make decisions based on what is
best for their infant. They are frequently overwhelmed and
feel that they simply must trust, rather than fight to be more
involved in decision making (Pinch, 1989,1990). However, this
lack of parental involvement and power also causes difficulties for nurses, because the nurses know that they (nurses)
will rock the boat if they give parents certain information
that might cause parents to question the medical treatment
plan. The potentially positive effect of suggesting that families call a meeting of the ICRC, through which their decisions may be affirmed, is only positive if the ICRC tends to
place great weight on parents' judgments about what is
best for the infant. Although nurses usually sit on the ICRC,
their influence varies dramatically from one institution to
the next.

Resolving Dilemmas
Within the Amendments' Framework:
Are Ethical Solutions Possible?

There needs to be more recognition that NICU treatment
can be not only burdensome, but also intolerably cruel for
some infants. Those who will have lives of very low quality
and whose lives are being sustained by high-tech NICU care
can certainly be seen as infants in need of advocates. Where

parents are not being given essential information about their infants' conditions and prognoses, we need to correct that situation. Where nurses' views of infants' responses to therapy are being rejected or discounted, the situation needs to be changed. The entire system of care in the NICU needs to become more humanized, with hospice approaches offered for infants who have little hope for a minimally decent recovery to health (Butler, 1986). Parents need more information, given in a highly sensitive, supportive manner, throughout their infant's hospital course. The values and judgments of families need to be respected when it is obvious that families are putting the infant's best interest ahead of all other considerations. Nursing administration needs to institute policies and procedures through which nursing input in decision making will receive the attention nurses' professional judgment deserves. Can these aims be accomplished within the framework of the amendments? Although I wish I could answer this question in the affirmative, I must admit that I cannot. The solution, unfortunately, may lie in reworking the amendments to allow greater decision-making prerogatives for parents. It seems that only through such reworking can we support parents who understand their infant's condition and prognosis, put the infant's welfare above all else, and choose to pursue or to forgo life-sustaining therapies based on their values about what kind of life can be a good life for their child.

References

Anspach, R. R. (1987) Prognostic conflict in life-and-death decisions: The organization as an ecology of knowledge. *J. Health Soc. Behav.* **28,** 215–231.

Baerts, W. and Meradji, M. (1985) Cranial ultrasound in preterm infants: Long-term follow up. *Arch. Dis. Child.* **60,** 702–705.

Bernbaum, J. C., Russell, P., Sheridan, P. H., Gewitz, M. H., Fox, W. W., and Peckham, G. (1984) Long-term follow-up of newborns with persistent pulmonary hypertension. *Crit. Care Med.* **12(7),** 579–583.

Berry, F. A. (1986) The anesthetic management of the premature nursery graduate, in *Anesthetic Management of Difficult and Routine Pediatric Patients* (Berry, F. A., ed.), Churchill Livingstone, New York.

Breshnahan, J. F. (1987) Suffering and dying under intensive care: Ethical disputes before the courts. *Criti. Care Nursing Q.* **10,** 11–16.

Butler, N. C. (1986) The NICU culture versus the hospice culture: Can they mix? *Neonatal Network* **5,** 35–42.

Butler, N. C. (1987) The ethical issues involved in the practice of surgery on unanesthetized infants. *AORN Journal* **46,** 1136–1142.

Butler, N. C. (1989) Infants, pain and what health care professionals should want to know: An issue of epistomology and ethics. *Bioethics* **3(3),** 181–199.

Calame, A., Fawer, C. L., Anderegg, A. and Perentes, E. (1985) Interaction between perinatal brain damage and processes of normal brain development. *Dev. Neurosci.* **7,** 1–11.

Cameron, M. (1986) The moral and ethical component of nurse burnout. *Critical Care Management Edition* **17(4),** 42B–42E.

Child Abuse Amendments of 1984, Pub. L. No. 98-457, 121, 98 Stat. 1749,1752.

Dennis, K. E. (1983) Nursing's power within the organization. What research has shown. *Nursing Aministration Quarterly* **8,** 47–57.

Duff, R. S. (1987) "Close-up" versus "distant" ethics: Deciding the care of infants with poor prognosis. *Semin. Perinatol.* **11(3),**244–253.

Escalona, S. K. (1982) Babies at double hazard: Early development of infants at biologic and social risk. *Pediatrics* **70,** 670–676.

Fleischman, A. R. (1986) An infant bioethical review committee in an urban medical center. *Hastings Cent. Rep.* **16(3),** 16–18.

Fleming, G. V., Hudd, S. S., LeBailly, S. A., and Greenstein, R. M. (1990) Infant care review committees: The response to federal guidelines. *Am. J. Dis. Child.* **144,** 778–781.

Franck, L. S. (1987) A national survey of the assessment and treatment of pain and agitation in the neonatal intensive care unit. *J. Obstet. Gynecol. Neonatal Nursing Nov-Dec* 387–393.

Gorski P. (1984) Experience following premature birth-stresses and opportunities for infants, parents, and professionals, in *Frontiers of Infant Psychiatry*, vol. 2 (Call, J, ed.), Basic Books, New York.

Gorski P., Davidson M. J., and Brazelton T. B. (1980) Stages of behavioral organization in the high risk neonate: Theoretical and clinical considerations. *Semin. Perinatol.* **3,** 61–72.

Gottfried A. W. and Gaiter J. L., eds. (1985) *Infant Stress Under Intensive Care* (University Park Press, Baltimore, MD).

Gramelspacher, G. P., Howell, J. D., and Young, M. J. (1986) Perceptions of ethical problems by nurses and doctors. *Archives of Inter-*

nal Medicine **146,** 577–578.

Greenspan S. and Porges S. W. (1984) Psychopathology in infancy and early childhood: Clinical perspectives on the organization of sensory and affective-thematic experience. *Child Devel.* **55,** 49–70.

Harrison, H. (1986) Neonatal intensive care: Parents' role in ethical decision making. *Birth* **13(3),** 165–175.

Jacobson, S. F. (1978) Stressful situations for neonatal intensive care nurses. The American Journal of Maternal-Child Nursing 3, May/June, 144–150.

Kraybill, E. H., Kennedy, C. A., Teplin, S. W., and Campbell, S. K. (1984) Survival, growth, and development of infants with birth weight less than 1001 grams. *Am. J. Dis. Child.*

Lawson, J. R. (1986) More on newborn surgery without anesthesia. *Birth* **15(1),** 36.

Lawson, J. R. (1988) Standards of practice and the pain of premature infants. *Zero to Three* **9,** 1–5.

McCormick., M. C. (1985) The contribution of low birth weight to infant mortality and childhood morbidity. *N. Engl. J. Med.* **312,** 82–90.

McCormick, R. A. (1974) To save or let die. *America.*

McMenamin, J. B., Shackelford, G. D., and Volpe, J. J. (1984) Outcome of neonatal intraventricular hemorrhage with periventricular echodense lesions. *Ann. Neurol.* **15(3),** 285–290.

Mitchell, C. (1983) New directions in nursing ethics. *Massachusetts Nurse* **50(7),** 7–10.

Penticuff, J. H. (1985) Reactions of the child and family to hospitalization in S. R. Mott, N. F. Fazekas, and S. R. James (eds.), *Nursing care of children and families: A holistic approach.* pp. 873–901. Addison-Wesley, Menlo Park, CA.

Penticuff, J. H. (1988) Neonatal intensive care: Parental prerogatives. *J. Perinat. Neonatal Nursing* **1(3),** 77–86.

Penticuff, J. H. (1989) Infant suffering and nurse advocacy in neonatal intensive care. *Nursing Clin. North Am.* **24(4),** 987–997.

Penticuff, J. H. (1990) Nurses' Ethical Decision Making in Perinatal Settings. Research grant #1 F33 NRO6472-01, National Center for Nursing Research, National Institutes of Health.

Pinch, W. J. (1989) Ethical decision making for high-risk infants: The parents' perspective. *Nursing Clinics of North America* **24(4),** 1017–1023.

Pinch, W. J. and Spielman, M. L. (1990) The parents' perspective. Ethical decision making in neonatal intensive care. *Journal of Advanced Nursing* **15,** 712–719.

Rice, B. R. and Feeg, V. D. (1985) First-year developmental outcomes for multiple-risk premature infants. *Pediatr. Nursing* 30–35.

Sameroff, A. J. (1981) Longitudinal studies of preterm infants, in *Preterm Birth and Psychological Development* (Friedman, S. L. and Sigmond, M., eds.), Academic, New York.

Scanlon, J. (1986) Barbarism. *Perinatal Press* **9,** 103,104.

Schraeder, B. D. (1986) Developmental progress in very low birthweight infants during the first year of life. *Nursing Res.* **35,** 237–242.

Schraeder, B. D., Rappaport, J., and Courtwright, L. (1987) Preschool development of very low birth weight infants. *Image* **19,** 174–177.

Stewart, A. L., Thornburn, R. J., Hope, P. L., Goldsmith, M., Lipscomb, A. P., and Reynolds, E. D. R. (1983) Ultrasound appearance of the brain in very preterm infants and neurodevelopmental outcome at 18 months of age. *Arch. Dis. Child.* **58,** 598–604.

TeKolste, K. A., Bennett, F. C., and Mack, L. A. (1985) Follow-up of infants receiving cranial ultrasound for intracranial hemorrhage. *Am. J. Dis. Child.* **139,** 299–303.

Walker, D. J. B., Feldman, A., Vohr, B. R., and Oh, W. (1984) Cost–benefit analysis of neonatal intensive care for infants weighing less than 1,000 grams at birth. *Pediatrics* **74,** 20–25.

Weir, R. F. (1984) *Selective Nontreatment of Handicapped Newborns: Moral Dilemmas in Neonatal Medicine* (Oxford University Press, New York), p. 52.

Wilkerson, J. M. (1987) Moral distress in nursing practice: Experience and effect. *Nursing Forum* **23(1),** 16–29.

Williamson, D., Desmond, M. M., Wilson, G. S., Murphy, A., Rozelle, J., and Garcia-Prats, J. A. (1983) Survival of low-birth-weight infants with neonatal intraventricular hemorrhage. *Am. J. Dis. Child.* **137,** 1181–1184.

Yu, V. Y. H., Downe, L., Astbury, J., and Bajuk, B. (1986) Perinatal factors and adverse outcome in extremely low birthweight infants. *Arch. Dis. Child.* **61,** 554–558.

Infant Care
Review Committees
in the Aftermath
of Baby Doe

Norman Fost

The public debate about treatment of handicapped newborns was dominated at first by consideration of substantive
ethical questions: Do some infants not meet accepted definitions of personhood? Why should euthanasia be prohibited
for infants who could permissibly be killed when they were
intrauterine? What role should quality of life play in treatment decisions? In part because of the difficulty in resolving
or gaining consensus on substantive questions, procedural
questions came to dominate the debate: Who should decide?
What limits, if any, should there be on parental discretion?

In the early 1970s, proposals were advanced advocating
the use of institutional committees in developing policies or
resolving difficult cases. Although these suggestions were not
initially limited to cases involving newborns, the growth of
committees created primarily to resolve the "Baby Doe" crisis provided the major impetus of the formation and use of

From: *Compelled Compassion* Eds.: Caplan, Blank, and Merrick
©1992 The Humana Press Inc.

such committees. This chapter will review the history of this movement, including its conceptual origins, review the ways in which these committees actually work, comment on some of the early concerns, and identify the current advantages and problems with such committees (Cranford, 1984).

History

The use of institutional committees to help resolve clinical ethical dilemmas did not arise with the Baby Doe controversy. Hospitals had previously used committees to help resolve dilemmas about sterilization and abortion (Levine, 1984). The use of a committee to decide which patients would have access to a limited number of dialysis machines evoked controversy because of their reliance on arbitrary criteria of social worth (Sanders and Dukeminier, 1968). In 1966, in response to evidence of widespread unethical research, the US Department of Health, Education, and Welfare (HEW) mandated local committees as the primary means of protecting human subjects of experimentation. Despite widespread criticism that these predominantly peer groups would not be effective, they eventually attracted widespread support and praise.

Suggestions to use committees to help resolve decisions about withholding and withdrawing life support from critically ill patients began in the early 1970s with suggestions by Veatch (1972), Teel (1975), Robertson and Fost (1976), and others (Rosner, 1985). The New Jersey Supreme Court, in its decision approving withdrawal of life support from Karen Quinlan (*In re Quinlan*, 1976) stipulated that such decisions *must* be approved by a hospital "ethics committee." This term has been widely criticized as a misnomer on the grounds that the court's intention was to require a neurologic consultation committee whose primary function was to confirm the prognosis of patients alleged to be in a vegetative state. Weisbard analyzed the opinion in more detail, and concluded that the court had clearly intended a broader purpose for such committees and that ethics was indeed part of their mission (Weisbard, unpublished).

Despite a growing literature on the subject throughout the 1970s, by 1983 a survey conducted for the President's Commission on Ethical Problems in Medicine discovered only 17 such committees in a cohort of 602 hospitals (Youngner et al., 1983). By 1985, two surveys showed that 50–60% of hospitals had formed such committees (McCarrick and Adams, 1989; Fleming et al., 1990). This extraordinary growth occurred on a voluntary basis. There were no statutes or regulations requiring such committees at the time. Even the influential final Baby Doe regulations, written in 1985 pursuant to amendments to the Federal Child Abuse Act, only recommended such committees (US Department of Health and Human Services [DHHS], 1985).

Motives and Purposes

Why did physicians voluntarily impose on themselves this opportunity for others to intrude on the doctor–patient relationship, and expose complex and controversial decisions to a broad group of individuals with no need to know? Several motives appear to have been important.

Many committees were formed primarily for educational purposes, either as discussion groups within a hospital, or facilitators for larger programs and conferences. From this relatively uncontroversial beginning, a consultation function would later emerge, as members of the group brought their own cases as food for thought and then prospective cases.

In some hospitals, administrators and risk managers sometimes were the driving force. Highly publicized cases, with or without involvement of the courts, were undesirable. Some committees were formed on an *ad hoc* basis to help resolve controversial cases that were threatening to spread beyond the hospital walls. More commonly, committees were formed to offer a forum for angry members of the hospital staff, particularly nurses, to at least ventilate and perhaps find allies to help avert what they perceived to be immoral management of difficult cases.

Separate from the desire to avoid adverse publicity or the perceived need to find a means to resolve political disputes within the hospital, some hospital attorneys perceived that institutional review could reduce the risk of lawsuits being initiated and possibly reduce the probability of suits ultimately succeeding. A charge of negligence—failure to take care—would be more difficult to sustain if the attending physician consulted with a broadly based group that made careful deliberations and kept detailed records. Similarly, a district attorney would be less likely to bring criminal charges against a physician who had broad institutional support.

Finally, there were some who believed and perhaps persuaded their institutions that such committees would improve the quality of ethical decisions and reduce the risk that decisions would in retrospect be perceived as unethical. An example of this view was the appeal to "ideal observer theory"—the claim that actions could be considered morally right only if they could be approved by an ideal ethical observer with the qualities of: omniscience—knowing the relevant and available facts; omnipercipience—the ability to vividly imagine the feelings of others involved; disinterest—having no vested interest in the outcome; dispassion—not being overwhelmed with emotion when critical decisions were being made; and consistency—deciding similar cases similarly. Since no human could possess these God-like qualities, the argument was made that properly formed and led committees could come closer to emulating this ideal (Robertson and Fost, 1986).

Form and Function

Despite the lack of mandated rules regarding the structure and function of committees, they generally seem to have developed along the general lines recommended by a task force of the American Academy of Pediatrics (American Academy of Pediatrics, 1984), the general outline of which was incorporated into the Baby Doe Regulations (US DHHS, 1985). This model suggested certain kinds of expertise—medical, nursing, social work, legal, and ethical—and representation of interested groups, including the patient and his family, nursing, hospital administration, and advocates for handicapped persons (Fost and Cranford, 1985).

There continues to be controversy about the advisability of the hospital attorney as a member of the committee. Although legal expertise is clearly relevant to complicated decisions regarding withdrawal of life-saving treatment, there has been persistent concern that such an individual might dominate the discussion, whether by expertise, personal style, or claims that risk management considerations should trump all other concerns. Such domination by legal claims is especially worrisome, since the law is so unclear and apparently tolerant in practice of almost any decision, no matter how egregious. Not a single physician in the US has ever been found civilly or criminally liable for withholding or withdrawing life support from any patient for any reason, yet hospital attorneys and physicians persist in the belief that this is an area of great risk of liability (Fost, 1989; Kopelman et al., 1988).

The need to obtain legal advice and avoid legal domination has been met in some committees by recruiting an academic or community lawyer not in the hospital's employ or by obtaining the hospital attorney's advice after the deliberations about the ethical issues. Obviously, many hospitals have the benefit of attorneys who perceive that doing what is best for the patient is the best legal advice and that defensive law is as inappropriate as defensive medicine (Weisbard, 1986).

Although committees have taken on a variety of functions, including educational activities and participating in development of hospital policies, it is the consultation function that has attracted the greatest interest, though this might not be the most important factor in a committee's impact on an institution, since only a small percentage of cases are reviewed. Access for consultation is typically available to members of the staff involved in the care of a patient or family members who are in disagreement with the attending physician. Relatively few institutions seem to have mandated review.

Committees vary in the aggressiveness with which they attempt to affect clinical decisions, ranging from simple discussion groups with no formal conclusions to some committees that actually claim authority to make decisions. The most common mode appears to be a consultation model, intended to develop consensus, which might be simply identified by the chair or put in the form of a recommendation, with or without a note in the chart, and with minutes of widely varying detail.

Problems

Many of the early concerns and criticisms of ethics committees have failed to materialize. Right-to-life advocates warned that they would do little more than create a patina of respectability, and continue to authorize withholding and withdrawing of treatment from infants with good prospects for meaningful life. In fact, the opposite seems to have happened. The two most common conditions in the controversial Baby Doe cases were Down syndrome (Annas, 1979; Steinbock, 1984; Weir, 1987) and spina bifida. Deaths resulting from withholding of standard treatment from infants with these diagnoses were common (Gustafson, 1973; Shaw et al., 1977; Will, 1982; Gross et al., 1983). This practice appears to have vanished since 1985. Despite a claim by the US Civil Rights Commission that there continued to be widespread abuse, the chairman of the commission refused to sign the report because of the lack of documentation (US Commission on Civil Rights [CCR], 1989).

Similar criticisms were made regarding the Institutional Review Boards (IRBs) when the US Department of Health Education and Welfare (HEW) (the forerunner of DHHS), in 1966, first proposed them as the major mechanism of protection for human subjects of experimentation. Although many of the initial committees were not providing adequate protection (Barber et al., 1973), by the time more comprehensive regulations were developed (US DHHS, 1983), there was considerable evidence and consensus that the incidence of seriously unethical research had been dramatically reduced, and that the IRBs were generally fulfilling their role in protecting subjects.

Some were concerned about practical problems, warning that a committee could not be assembled in the middle of the night to decide whether to resuscitate a premature infant. In fact, the vast majority of decisions regarding withholding or withdrawing treatment are made under circumstances that allow considerable time for reflection, and the occasional true ethical emergencies, such as whether to resuscitate a very small premature infant in the delivery room, can be addressed through hospital policies.

The most serious persistent concern about ethics committees has been the opposite of the fear of rubber-stamping inappropriate undertreatment, namely, that they would instead become tools of risk managers, and develop a bias toward overtreatment (Weir, 1987). There is evidence that the original Baby Doe problem—inappropriate undertreatment of infants with good prospects for meaningful life—has been replaced by a higher incidence of overtreatment, meaning maintenance of biologic existence of children with little or no prospects of intact survival or a life that is likely to bring any pleasure to the child (Kopelman et al., 1988; Pomerance et al., 1988).

Advantages of Committees

Those who advocated ethics committees as a political compromise between the *status quo ante* and more Draconian alternatives, such as the dreaded federal hot line and "flying Baby Doe squads," have achieved their goal. Despite occasional criticism that more regulation is needed (US CCR, 1989), the controversy about management of handicapped infants seems to have subsided.

Those concerned with the substantive problem of undertreatment of infants with Down syndrome, spina bifida, and other conditions compatible with a long, happy life also should be satisfied that the mission has been accomplished. For those looking for more thoughtful, informed, and considered decisions regarding infants with more ambiguous futures, committees have provided considerable support in many institutions. There has not been a comprehensive survey since the 1985 study conducted by Fleming et al. (Fleming et al., 1990), but my personal experience, reports at national conferences, and other contacts support a conservative estimate that thousands of cases at hundreds of hospitals have been reviewed by such committees in the past five years. A review of approximately 100 cases in our own experience can be classified in three categories.

Supportive

The most common reason for consultation has been a difficult case in which there is no disagreement about how to proceed, but in which the medical and/or nursing staff seeks confirmation that a proposed course of action does not violate ethical or legal norms. An example is illustrative:

> 6 month old male, hospitalized since birth, now being treated for chronic renal failure, severe bronchopulmonary dysplasia, severe hydrocephalus secondary to Grade-IV intraventricular hemmorhage with recurrent shunt infections, and congestive heart failure secondary to complex congenital heart disease which is probably inoperable. The parents persistently requested that treatment be stopped. The attending physician agreed that the infant had almost no chance of surviving outside the nursery, virtually no prospect for social interaction, and was highly likely to die within 6 months, but he believed that the Baby Doe regulations required continued treatment. A nurse suggested consultation with the committee, which unanimously thought continued treatment was not in the

child's interest. The physician was relieved to have the support, asked that a note be put in the chart, and discontinued the ventilator following which the child died.

Clarify Facts

The most common reason for disagreements resulting in consultation has been confusion about empiric questions, medical or legal. The Committee process often clarifies the facts, which reveals consensus about how to proceed.

3 month old male with apparent central hypoventilation ["Ondine's curse"], but otherwise healthy and apparently normal neurologically. The parents have asked that ventilator support be discontinued, stating they do not want the child in their home if he is ventilator dependent. The staff are divided, but all are uncomfortable about discontinuation of treatment.

The committee was consulted by the attending physician. The consultation stimulated a broader search for information on longterm outcome of patients with this condition, showing that some outgrow their ventilator dependency. The committee also informed the parents that a medical foster home could probably be found, either for respite care or longterm and possibly permanent placement. The possibility of this "escape," along with the new information about prognosis, caused the parents to revise their thinking and they agreed to continue with treatment for the time being.

New Ideas

Many heads are better than one, and as the Committee accumulates experience in managing difficult cases it is often able to suggest a new approach that leads to consensus.

A 9 year old vegetatively retarded boy is permanently institutionalized in a state institution. He does not seem to have any social interactions. He has recurrent pneumonia for which he is transferred to an acute care hospital, frequently for a prolonged admission. The parents do not want him transferred, believing his life is filled with suffering with little compensating pleasure. The medical staff do not disagree, but the institutions believe they are required to provide all medically beneficial treatment according to the "Baby Doe" regulations. They are also under some political pressure.

The parents consult the Committee at the acute care hospital, expecting them to refuse to admit their son when he becomes acutely ill. The suggestion is made that the parents take their child home when he is ill. Neither institution has any objection. The child subsequently died at home after a short illness.

Conclusions

Hospital ethics committees have become a growth industry over the past decade, for a variety of reasons, but without legal requirement. This growth has been accompanied by a virtual disappearance of the "Baby Doe" problem—withholding medically beneficial treatment from infants who had excellent prospects for long, meaningful lives. Concurrently, there has been an apparent increase in overtreatment—administration of invasive treatment to maintain the lives of infants who appear to have little or no prospects for meaningful lives. Ethics committees have been helpful in preventing some cases in both categories.

Committees do not typically work by explicit analysis or application of ethical issues in the traditional academic sense (Annas, 1984). Rather, they seem to operate primarily as consensus development forums, helping to clarify or resolve misunderstandings about facts, including prognosis and alternative care arrangements. Often, they simply ratify or support a view that is not controversial. Sometimes, probably infrequently, they play a major role in reversing a decision that would be widely perceived as ethically problematic.

In some institutions, ethics consultants have become a more acceptable, efficient, and less intrusive resource for resolving similar problems. Sometimes these consultants operate independently, but more commonly they are either emissaries from or closely linked to an institutional committee. Within states and regions, networks of ethics committees are also on the rise (Kushner, 1989; Sagin, 1989) and there is even a budding "network network"—an organization of ethics committee networks.

Ethics committees seem to be entrenched and are rapidly becoming as accepted as the institutional review boards that review research on human subjects. In many institutions, they are even more respected than the IRBs, particularly because they are voluntary and their consultations are at the request of physicians, rather than required by law.

Whether or not they are doing a good job depends in part on a definition of their role and purpose, and on more data than are currently available (Mahowald). The next decade will undoubtedly witness expanded self-reporting of their activities and experiences for public review, as well as more formal studies of their behaviors and norms.

References

American Academy of Pediatrics, Infant Bioethics Task Force and Consultants (1984) Guidelines for infant bioethics committees. *Pediatrics* **74(2)**, 306–310.

Annas, G. J. (1979) Denying the rights of the retarded: The Philip Becker case. *Hastings Cent. Rep.* **9**, 18.

Annas, G. J. (1984) Ethics committees in neonatal care: Substantive protection or procedural diversion? *Am. J. Public Health* **74**, 843–845.

Barber, B., Lally, J. J., Makarushka, J. L., and Sullivan, D. (1973) *Research On Human Subjects* (Russell Sage, NY).

Cranford, R. E. and Doudera, A. E., eds. (1984) *Institutional Ethics Committees and Health Care Decision Making* (Health Adm. Press, Ann Arbor, MI).

Fleming, G., Hudd, S. S., LeBailly, S. A., and Greenstein, R. M. (1990) Infants care review committees: The responses to federal guidelines. *Am. J. Dis. Child.* **144**, 778–781.

Fost, N. (1989) Do the right thing: Samuel Linares and defensive law. *Law Med. Health Care* **17(4),** 330–334.

Fost, N. C. and Cranford, R. (1985) Hospital ethics committees: Administrative aspects. *JAMA* **253,** 2687.

Gross, R. H. et al. (1983) Early management and decision-making for the treatment of myelomeningocoele: A critique. *Pediatrics* **73,** 564–566.

Gustafson, J. (1973) Mongolism, parental desires and the right to life. *Persp. Biol. Med.* **16,** L529–L557.

In re Quinlan (1976) 70 NJ 10, 355 A2d 647, Cert Denied 429 US 922.

Kopelman, L. M., Irons, T. G., and Kopelman, A. E. (1988) Neonatologists judge the Baby Doe rule. *N. Engl. J. Med.* **318,** 677–683.

Kushner T. (1989) Networks across America. *Hastings Cent. Rep.* **19(1),** 24.

Levine, C. (1984) Questions and answers about hospital ethics committees. *Hastings Cent. Rep.* **14,** 9–12.

McCarrick, P. M. and Adams, J. (1989) *Ethics Committees in Hospitals.* Scope Note 3 (Kennedy Institute of Ethics, Georgetown University, Washington, DC).

Mahowald, M. B. (1989) Baby Doe committees: A critical evaluation. *Clin. Perinatol.* **15(4),** 789–800.

Pomerance, J. J., Yu, T., and Brown, S. J. (1988) Changing attitudes of neonatologists toward ventilator support. *J. Perinatol.* **8:3,** 232–241.

Robertson, J. A. and Fost, N. (1976) Passive euthanasia of defective newborn infants: Legal considerations. *J. Pediatrics* **88,** 883–889.

Rosner, F. (1985) Hospital medical ethics committees: A review of their development. *JAMA* **253(18),** 2693–2697.

Sagin T. (1989) The Philadelphia Story. *Hastings Cent. Rep.* **19(1),** 24.

Sanders, D. and Dukeminier, J. (1968) Medical advance and legal lag: Hemodialysis and kidney transplantation. *UCLA Law Rev.* **15,** 366–380.

Shaw, A., Randolph, J. G., and Manard, B. (1977) Ethical issues in pediatric surgery: A national survey of pediatricians and pediatric surgeons. *Pediatrics* **60,** 588–599.

Steinbock, B. (1984) Baby Jane Doe in the courts. *Hastings Cent. Rep.* **14(1),** 13–19.

Teel, K. (1975) The physician's dilemma—a doctor's view. *Baylor Law Rev.* **27,** 609.

US Commission on Civil Rights (1989) *Medical Discrimination Against Children with Disabilities.* US Government Printing Office.

US Department of Health and Human Services (1983) *Regulations on Protection of Human Subjects.* 45 CFR 46.

US Department of Health and Human Services (1985) Service and Treatment for Disabled Infants: Model Guidelines for Health Care Providers to Establish Infant Care Review Committees. 50 *Fed. Register* 14,893–14,901.

Veatch, R. (1972) Choosing not to prolong dying. *Med. Dimensions* **40,** 8–10.

Weir, R. F. (1987) Pediatrics ethics committees: Ethical advisors or legal watchdogs? *Law Med. Health Care* **15(3),** 99–108.

Weisbard, A. (1986) Defensive law: A new perspective on informed consent. *Arch. Int. Med.* **146,** 860–861.

Weisbard, A. What role for institutional committees: Ethics or prognosis? (unpublished manuscript).

Will, G. W. (1982) The killing will not stop. *Washington Post.*

Youngner, S. J., Jackson, D. L., Coulton, C., Juknialis, B. W., and Smith, E. (1983) A national survey of hospital ethics committees, in President's Commission for the Study of Ethical Problems in Biomedical and Behavioral Research. *Deciding to Forego Life-Sustaining Treatment*, Appendix F, US Government Printing Office, Washington, DC, pp. 443–449.

Decision Making in the Neonatal Intensive Care Unit

The Impact of the 1984 Child Abuse Amendments

Terry Walman

The focus of this chapter is on the impact of the 1984 Child Abuse Amendments on government institutions, the involved interest groups, and affected families and infants. The commentary provoked by the amendments ranges across the political and philosophical spectra. The ambiguous, carefully wrought language of the legislation allowed spokespersons for each special-interest group to claim victory for their side. The president of the American Academy of Pediatrics was quoted in a press release as saying: "It would appear the final rule reaffirms the role of reasonable medical judgement and that decisions should be made in the best interests of the infant" (Murray, 1985). About the same time, the National Right to Life Committee general counsel was quoted as saying, "We have come a long way in three years, from the death of Infant Doe to the passage of historic legislation by Congress, and now the release of these HHS regulations to enforce that legislation" (*NRLN*, 1985).

From: *Compelled Compassion* Eds.: Caplan, Blank, and Merrick
©1992 The Humana Press Inc.

However, while claiming victory in the battle over the Child Abuse Amendments passed by Congress, it is also clear that each side was dissatisfied with the result imposed by the legislation. A special article in one of the respected general medical journals published the results of a survey answered by 49% of the neonatologist members of the American Academy of Pediatrics on whether the Child Abuse Amendments had affected their practices. Of the 494 respondents, 76% believed the Child Abuse Amendments were not necessary to protect the rights of handicapped infants, and 60% believed that the regulations did not allow adequate consideration of infants' suffering. Additionally, 56% agreed that infants with extremely poor prognoses for survival were being *overtreated*, and 33% complained that they had altered their practice and treatment decisions as a result of the new federal regulations (Kopelman et al., 1988).

Although the physician-neonatologist group called for a reevaluation of the Child Abuse Amendments by Congress or the courts because they mandated unjustified overtreatment of this category of newborns and, thus, caused "unnecessary pain to infants and their families," (Kopelman et al., 1988), Professor Patricia Caulfield of Valparaiso University Law School complained that handicapped infants who fell into any of the three exceptions outlined by the Final Rule where withholding of treatment would not be considered medical neglect (other than *appropriate* nutrition, hydration, or medication), were not protected by binding enforcement procedures from physicians who would elect to withhold treatment, which would result in the infants' deaths. "The Amendments imperil the life of the newborn" (Caulfield, 1986). Thus, Caulfield, voicing a vitalist viewpoint, called for Congress to further legislate constitutional due process rights to protect the handicapped infant from treatment denial (Caulfield, 1986).

Is there any way to reconcile this enormous difference of opinion between the Vitalist position and the Neonatologists surveyed? I think it is fair to point out that neither of these groups is particularly satisfied with the current impact of the Child Abuse Amendments, because no one group gained very much from the compromise legislative effort that produced these federal regulations. No one group gained very much, because no group had to give up very much. The whole issue of "difficult treatment decisions concerning handicapped children" became unavoidably a political issue when the focus turned away from the courts and was taken up by Congress in the summer of 1984 (*Bowen v. Am. Hosp. Assoc.*, 1982). This is not to say that the debate prior to that time had not been politically motivated. The one clear winner emerging from the Baby Doe controversy was the Reagan Administration, which gained political prestige with both the right-to-life proponents and advocacy groups for the handicapped. While claiming the moral high ground of protecting civil liberties by guarding the interests of newborns and protecting the rights of the handicapped, the Reagan Administration continued to cut federal aid for many programs that had been enacted to assist the handicapped (Lantos, 1987). This was political exploitation at its most cynical plane.

Congress made out slightly less well than the executive branch, because it recognized the irreconcilable differences between the different interest groups and arranged a compromise solution that each could claim as a victory. The satisfaction that each interest group experienced has been predictably short-lived as the hollowness of its victory has become more apparent. Such is the nature of political expediency—especially when the solution is as purposely vague and ambiguous as the Child Abuse Amendments were intentionally designed to be. The confusion experienced by each side is expressed in their widely divergent viewpoints of what the law should require and demand for legislative reform.

The third branch of the federal government—the judiciary—came out essentially neutral in the end. When the forum for debate in this controversy left the courts and entered the halls of Congress, the Supreme Court was taken out of the loop. The judiciary fulfilled its role by focusing the arguments against the use of Section 504 as a legitimate basis for federal intervention in this area. The ultimately unremarkable plurality opinion generated by *Bowen v. Am. Hosp. Ass'n.* does not give any indication of how the Supreme Court might deal with the federal governmental intervention allowed by the Child Abuse Amendments anchored in the state interest of child neglect.

Pediatricians did not fare quite as well as the winners and nonlosers in politics and government (respectively). The pediatricians seemed to be in a no-win situation. On the one hand, they achieved an implicit and timely recognition of their skills as medical technicians or miracle workers in the neonatal intensive care unit (NICU) (Berseth, 1987), but at the same time, they ceded a bit of their moral authority since their role as child advocates was challenged (Lantos, 1987). The responses of the neonatologists in the Kopelman study on whether the regulations had affected their practices indicates that, as a group, these neonatologists felt as though they had lost some of their professional integrity as a result of the enactment of the Child Abuse Amendments. The neonatologists would also insist that "Baby Doe" infants were worse off than before, because they were more frequently subjected to inappropriate overtreatment amounting to unjustified infliction of pain and suffering not in their (the infants') best interest.

The right-to-life advocates must be viewed as at least partial winners in the promulgation of the Child Abuse Amendments, because they gained a great deal of recognition by the very nature of the national debate that ensued. It was their initiative that called the Reagan Administration to the forefront of the battle, and even if they achieved less than a satisfactory result, their proponents readily admit that they are further along their agenda than when they started (Caufield, 1986).

The advocates for the handicapped cannot claim such an improvement in status. Although they were able to call attention to their cause with the promulgation of the Section 504 regulations by the Dapartment of Health and Human Services (DHHS), the real focus of the Rehabilitation Act of

1973 (and Section 504) was *distorted* by the attempts of the federal government to intervene in individual medical decision making as it pursued the Baby Doe and Baby Jane Doe litigation. The ultimate outcome was a reaffirmation of the *real* meaning of Section 504 (by the courts) as it pertains to handicapped adult *opportunities*; however, in spite of this recognition, there was a loss of funding for many of the federal programs that provided the vocational and educational services to create those opportunities. In the end, the attention paid to their cause was short-lived, and thereafter, the handicapped were cast aside—worse off than when they began.

Clearly, however, the group that suffered the worst outcome and lost the most prestige, respect, and authority was the parents of Baby Doe infants. Their capabilities and values were at the center of the controversy, but they were the least organized interest group and therefore were not effectively represented when the Child Abuse Amendments were considered. In spite of the rulings of the lower courts in all of the litigation reviewed here and the dicta of the Supreme Court plurality opinion in *Bowen v. Am. Hosp. Assn.*, the upshot of the Child Abuse Amendments and their accompanying regulations currently in effect is that:

> States accepting funds under the Act *must* require hospitals to *report* certain failures to treat. Situations requiring hospital reporting may include both those where parents refuse consent [and the unlikely instances where parents request treatment and the hospital declines to provide it]. The underlying logic of the federal reporting standard is that parents who deny permission for what the physician reasonably sees as required treatment [including presumably, long-term intravenous nutrients, water, and medication] *may* be guilty of child neglect or abuse. Lack of parental consent, so crucial to the outcome in *American Hospital Assn.* is irrelevant under the current federal definition of medical neglect of disabled infants

with life-threatening conditions. Such presumed medical neglect must be reported *regardless of whether
the parent refused consent"* (emphasis in original)
(Huefner, 1986).

The Child Abuse Amendments regulations ignore the
traditional role of parental consent in treatment decisions.
Although parental discretion is not completely ignored, as
was the original intent of the DHHS regulations promulgated
under Section 504, that discretion is now statutorily relegated
to second-class status behind the treating physicians' views
as to what is "medically indicated" treatment, "adequate"
medical care, or "appropriate" nutrition/hydration/medica-
tion. Furthermore, since most medical care in tertiary hospitals for critically ill newborns is carried out utilizing a medical
team rather than a single physician, does this mean that *any*
physician member of the hospital care team covering the
NICU can overrule the family's decision not to opt for a particular treatment? Could this be extended to resident physicians in training at the institution or even medical student
doctors peripherally involved in the infant's care? The more
people that are available to render their opinions in these
matters, the more diluted the concept of parental consent and
discretion becomes, especially where the choices and alternatives are in doubt and when reasonable and prudent medical judgment is not settled among the experts themselves.
      This scenario is not at all in keeping with the holding of
the Supreme Court in *Bowen v. Am. Hosp. Ass'n.* that, when
there is a choice among treatments or disagreement among
experts about what treatment is best, the *parents* ought to
be able to decide what is best for the child. "In broad outline,
state law vests decisional responsibility in the parents, in
the first instance, subject to review in *exceptional* cases by
the state acting as *parens patriae*" (emphasis on "exceptional"
added) (*Bowen v. Am. Hosp. Ass'n.*). Also, the current scenario is not consistent with the recommendations of the
*Report of the President's Commission for the Study of Ethi-
cal Problems in Biomedical and Behavioral Research:*

Deciding to Forego Life-Sustaining Treatment: "Parents should be the surrogates for seriously ill newborns unless they are disqualified by decision-making incapacity, an unresolvable disagreement between them, or their choice of a course of action that is *clearly* against the infant's best interest" (emphasis added) (President's Commission, 1983). Indeed, the Kopelman study found that nearly two-thirds of the respondents thought the current regulations affected parental rights to consent to or refuse treatment on the basis of what they thought was in their infant's best interest (Kopelman et al., 1988).

Thus, it is clear that parents and the privacy interests they may have enjoyed traditionally in their discretion to exercise meaningful consent (implicit in the concept of meaningful consent is the choice to be able to withhold that consent and to have that decision respected by those who sought the consent in the first instance [with regard to treatment decisions for their children]) were the real losers in the implementation of the Child Abuse Amendments. Whereas there was already precedent for overriding parental discretion in matters of child abuse and neglect outside the medical decision-making arena (and within the medical decision-making arena for older children), the inclusion of medical decision making, in the already tragic and difficult context of critically ill defective newborns under the guise of medical neglect, represents a further significant erosion of parental autonomy.

I was moved to become involved with the topic of the Child Abuse Amendments because of an experience I had recently as member of a teaching hospital's Patient Care Advisory Committee (a.k.a. "Ethics Committee"). These forums are actually mandated by statue in Maryland (ACM) and are a direct result of the voluntary Infant Care Review Committees proposed by the earlier Baby Doe Regulations.

In 1988, a young professional couple gave birth to an infant not unlike the original Baby Doe of Indiana. The mother had experienced a completely uneventful pregnancy; her age and past medical history did not suggest the need for amniocentesis. The birth of a Down syndrome baby (Trisomy 21) to them was unexpected and unwanted. The baby had several of the congenital abnormalities associated with Down syndrome, and required surgical intervention to correct esophageal atresia and tracheoesophageal fistula. (Esophageal atresia is the incomplete formation of the upper esophagus such that the mouth ends in a blind pouch. The distal esophagus is joined abnormally with the trachea [windpipe]. Thus, there can be no successful oral feeding, and additionally, the newborn's lungs are harmed by the continuity with the stomach; aspiration of stomach acids/enzymes is likely and leads to pneumonia. The condition is incompatible with life. With surgical correction of these lesions, the infant would be able to eat normally [oral], and the lungs would no longer be exposed to food or saliva. The underlying Down syndrome [Trisomy 21] would, of course, not be affected). The physicians attending the newborn sought consent for the operation from the parents, but after a series of sincere in-depth and informed discussions with a number of physicians, nurses, clergy, and members of their family, the parents exercised their right to withhold consent and refused to authorize the proposed surgery. The scenario to this point was very similar to that of the "Johns Hopkins baby" case in 1972, when parental refusal to authorize surgical intervention to correct duodenal atresia in a newborn with Down syndrome was respected, and the infant died in the infant nursery.

However, this was 1988, and the Child Abuse Amendments of 1984 had an impact on what was to happen subsequently. The parents were told that, if they would not consent to the surgery, the hospital was obligated to report the case to the state Child Protection Services agency, which would investigate whether the newborn was a victim of "medical neglect." If the state authorities found the child to be such a victim, the state would intervene under "*parens patriae*" proceedings to take over guardianship for purposes of authorizing the tracheal and esophageal surgery. All of this was explained by the treating physicians in such a fashion as to encourage the parents to change their minds and authorize the surgical treatment plan.

Most new parents in this situation would probably have yielded to the threat of state intervention, especially upon being accused of "medical neglect." However, these parents sought outside legal advice and returned to the hospital with legal counsel for further discussions about the care of their newborn. Ultimately, the parents chose *not* to authorize surgery, and as a result of the Child Protection Services intervention, they voluntarily gave up all claims and custody of the newborn to the state. The parents left the hospital without the child, and the infant became a ward of the state. Surgery was performed under guardianship consent. The parents were no longer involved because of the Child Abuse Amendments.

Other chapters in this book have discussed in some detail the history of the Baby Doe controversy and the developments that led to the federal infant Doe rule contained in the Child Abuse Amendments of 1984. The rule mandates aggressive intervention and treatment of severely deformed newborns with few exceptions, and essentially eliminates parental discretion and authority for treatment decisions whenever there is a disagreement with the treating physicians' opinions as to what is "medically indicated" treatment. "Medically indicated" treatment is a terribly subjective phrase that is intentionally ambiguous and likely to be interpreted differently by even the most reasonable of pediatric practitioners. Given the propensity for the team treatment concept prevalent in most tertiary hospitals where critically ill newborns are likely to be cared for, the practical effect of the Child Abuse Amendments is to privilege the view of those team members who hold the most aggressive treatment philosophy, even though they may not be the attending physician to the newborn patient. Indeed, the person who wishes to pursue the most aggressive treatment plan may not be the infant's primary physician and may not have entered into a physician–patient surrogate relationship with the parent.

In the idealized physician–patient relationship, the physician makes a diagnostic prognosis with regard to the various treatment options that are available. After becoming educated from this interaction with the physician as to the underlying medical condition and the alternatives to treatment as well as the risks and benefits of each treatment, the patient is able to make an informed choice to either accept or forgo a given treatment. When the patient is a child, the physician–patient relationship and the requirement of an informed consent to proposed medical therapy is the same, except that the parent becomes the surrogate decision maker for the child, who has neither the capacity to comprehend the information supplied regarding medical condition/prognosis/ treatment options, nor the ability to act autonomously in consenting to or forgoing medical treatment in his or her own best interest. The ethical and legal precedent for parental surrogacy in their children's medical care decision making is reflected in the President's Commission report of 1983, and the state and federal judicial decisions at virtually every level evoked by the Baby Jane Doe controversy.

In the unusual circumstances in which proposed medical treatment options are clearly beneficial to a minor child, and parental consent is refused such that medical treatment would be withheld lacking legal surrogate authorization, the state has the power to intervene, acting *"en parens patriae"* on behalf of the child's best interest to authorize the proposed medical treatment. This exception to override parental discretion is equally well-established.

If we were to analyze the medical decision-making authority vested in parents of seriously ill newborns in relation to physician assessment and recommendations for treatment, it would appear that the major effect of the Child Abuse Amendments of 1984 is to remove the parental discretion to forgo medical therapy when that proposed therapy is anything less than "futile or virtually futile in terms of survival of the infant *and* the treatment itself under such circumstances would be inhumane" (emphasis added) (USC, 1984).

I contend that, even though the Child Abuse Amendments of 1984 lack the binding qualifying definitions of the earlier Baby Doe rules, and in spite of the fact that they are grounded in the child abuse and neglect statues, rather than Section 504, their *impact* on parents whose infant's care team includes at least one physician who holds to the vitalist philosophy is still potentially devastating and abusive; worse, it is arbitrary and capricious. When a newborn is diagnosed to be critically ill and seriously impaired, the parents are extremely unlikely to be in a position to choose either a tertiary medical center or an infant care team prior to agreeing to the infant's initial care plan. (Recall that, in the case of Infant Baby Jane Doe discussed above, the severely impaired newborn was "immediately transferred to University Hospital. . . ." This is the typical scenario of tertiary care center admissions.) Only after a full assessment by the neonatologists at such a tertiary center are the parents in a position to make an informed decision about proposed medical intervention. By that time, it may be too late to stop compulsory treatment of the newborn set in motion by state intervention authorized under the Child Abuse Amendments. All that is needed is a single physician in the care team who clings to the vitalist philosophy. The privilege granted to physicians who would recommend aggressive medical intervention in all ambiguous or uncertain cases, where it would not be stated that therapy would be futile or virtually futile *and* humane, draws a new bright line for state intervention in a previously *gray* area where parental discretion was sought and respected.

Not only is it arbitrary that an infant care team physician might wish to pursue a vitalist philosophy in spite of parental disagreement, but there is little consistency as to which states have chosen to continue participation in the Child Abuse Prevention and Treatment Act Program (CAPTA) since the Child Abuse Amendments were added to the program's participation requirements. (As noted above, the Child Abuse Amendments of 1984, which gave rise to the current federal Infant Doe regulations, are conditions attached to the federal CAPTA program, which makes funds available to states for their child abuse and neglect agencies. States' participation is voluntary.) For example, California, Pennsylvania, and Indiana have elected not to participate in the CAPTA grant program (House Select Committee, 1987). Other states that do participate in CAPTA funding may decide to opt out in the future. In a survey by DHHS in 1987, 11 state child protective agencies responded that, because of the medical and ethical issues involved, their responsibility for Infant Doe cases may not be appropriate (Office of Inspector General, 1978). Therefore, state participation in CAPTA and, thus, the intervention by their child abuse/neglect agencies are not uniform.

Furthermore, it is not established either in medical practice/fact or in law which infant conditions or diagnostic categories fit within the key provisions of the Child Abuse Amendments of 1984 (Office of Inspector General, 1978). What is meant by "virtually futile treatment" in terms of survival of the infant with the additional requirement that "the treatment itself under such circumstances would be inhumane"? Would the Down syndrome Infant Doe case that occurred in Indiana meet this criterion? Presumably not, because Down syndrome infants, in spite of the frequency of congenital anomalies associated with the trisomy condition, may exhibit a wide range of morbidity and mortality outcomes.

What about another autosomal trisomy, such as "Trisomy 18"—a condition associated with severe anatomic malformations that is uniformly fatal in the first three months of life? Would treatment here "merely prolong dying" or otherwise be futile in terms of the survival of the infant? If Trisomy 18 fits into the exceptions allowing respect for parental discretion to withhold medical treatment even if proposed by a vitalist physician, then what about "Trisomy 13," another genetic anomaly associated with severe anatomic malformations and developmental impairments, but which is less predictable than Trisomy 18, and has an average life expectancy of about one year?

There are differing opinions about what the Child Abuse Amendments require. For example, Stephen Newman of New York Law School argues that "DHHS has even gone on record with the suggestion that under the statue, infants suffering from Tay-Sachs disease [A Tay-Sachs baby may appear normal at birth, but within a few months, the inherited metabolic derangement begins to cause muscle deterioration such that paralysis, spasticity, difficulty swallowing, deafness, blindness, and convulsions eventually develop. By age two or three, these infants become virtually vegetative, and they rarely live beyond age four to five. There is no curative treatment] plus a life-threatening intestinal blockage, must undergo intestinal surgery. The surgery would repair the blockage, enabling such infants to live for perhaps a year or two until the inevitable, prolonged and agonizing death of Tay-Sachs ends their lives" (Newman, 1989). Any treatment team physician who shared the DHHS view in a state that participates in the CAPTA program could presumably prevail as the *most aggressive treating physician* in forcing government intervention and treatment over the "medical neglect" of parents' preference to forgo such a treatment option. In contrast, Nancy Rhoden reported that the Department of Pediatrics at Yale University School of Medicine held a distinctly different opinion in this particular scenario: "Such a compulsory treatment requirement for this infant with Tay-Sachs should be viewed, and would be viewed by medical opinion and loving parents as an act of abuse and *inhumanity*" (emphasis added) (Rhoden, 1985). Clearly, a treatment that would be considered inhumane in New Haven, Connecticut, ought not to be compulsory in another jurisdiction, simply because it does not meet the definition set up in the current Child Abuse Amendments requiring inhumanity *and* virtually no chance of saving the life of the infant. Such a distinction is arbitrary, irrational, and too subjective for the kind of governmental intrusion into a fundamental area of otherwise private decision making.

In addition to the possible arbitrary and capricious intervention by the state into the very sensitive area of individual parental medical decision making described above, what becomes of the seriously impaired critically ill newborn whose immediate life-threatening defect is repaired by aggressive medical intervention over the refusal of parental consent? There are several possibilities: Either the guardianship of the state child neglect/abuse authorities is *temporary* and limited to the medical decision making at issue in our scenario, and the parents/family of the severely impaired newborn infant accept responsibility for the child's continuing care and provide a home for him or her, *or* as in the unfortunate case presented to the Ethics Committee in 1988 outlined above, the parents respond to the scenario of state interference into their discretionary refusal to authorize aggressive medical care by completely withdrawing from all future involvement with the baby's care and development, thus extending the state's temporary guardianship intervention into a more permanent responsibility for custodial care. By relinquishing all control and rights in the infant's future, the parents essentially abandon the child to the responsibility of the state. I do not wish to pass judgment on the parents who chose this course of action, but I would like to note several observations:

1. Such an option must be available for parents who find the result of such state interference to be unacceptable to them (whatever their reasoning);
2. The parents will likely suffer social and emotional turmoil as a result of their choice to "abandon" the infant to the state authorities; and
3. After "abandonment" to the state, the best interest of the child should be the focus of the state's efforts to provide care for the infant who has been the beneficiary of its *"parens patriae"* proceedings overriding the parental refusal of aggressive medical intervention.

It is not the purpose of this discussion to argue what the best interests of an infant in these circumstances would be, but I think it reasonable to contend that the best interest of any newborn would be to gain placement into a family environment wherein the neonate would be cared for in a loving and individualized way by people with whom the child could establish a stable and secure interactive relationship. Such an environment would be possible if the child could be adopted into a family that wished to replace the natural parents in a long-term/permanent relationship, where the child could develop its full interactive potential with the adoptive parents in the same way that natural parents ordinarily provide for their children. If permanent adoption was not immediately available, then foster home care or an institutional care setting with the same positive factors that would promote the infant's best interest, would be in order.

In the example of the 1988 infant case described above, after successful surgery to repair the esophageal blockage and the tracheoesophageal fistula, the infant was able to eat normally and develop with its underlying genetic condition of Down syndrome. Since the natural parents opted to surrender future responsibility for the infant to the state's child abuse/neglect protection services, it became the duty of the state to provide a proper home for the infant. However, there was not an adoptive home available for that particular infant. In fact, the infant had no place to go when it was medically ready for discharge, so she was forced to stay in the hospital long past the time when a similarly situated infant whose natural parents retained their custody rights would have been taken home postoperatively. Furthermore, when that severely impaired infant was finally discharged from the hospital, it was to a series of temporary foster homes and, ultimately, to a state institution where she remains today receiving custodial care for her impaired condition.

It is a sad commentary that "in the world of adoption, where healthy white infants are hotly pursued, a burgeoning group of 'special needs' children is left behind" (*Time*, 1987). The severely impaired newborn infants that survive as a result of aggressive medical intervention are clearly "special needs" children, and when they are "abandoned" to the custody of the state, they are not likely to be adopted. In fact, the more severe an infant's deficit, the *less* likely it is to be adopted. Also, the more severe an infant's deficit, the more difficult it is to care for him or her in foster homes or institutions. Without individualized care, it is less likely that such an infant will attain his or her own full measure of development.

Given the fact that state intervention to override parental refusal to aggressive medical therapy in the difficult setting of critically ill severely deformed newborns is promoted by the current federal Child Abuse Amendments and that such intervention may be arbitrary and irrational to the interest of the natural parents, the outcome of such government intervention is always an invasion of a privacy interest that is fundamental to parents/families. It makes little sense to invade the privacy of parental discretion in medical decision making with no greater goal than to sustain life for its own sake. If the vitalist philosophy reflected in the federal Child Abuse Amendments can be put into effect by the arbitrary conditions noted in the earlier discussion, then there must be some rational purpose to this exercise of governmental power. To merely prolong life in an infant who has special needs that will thereafter not be met is a hollow and purposeless exercise of governmental authority. I contend that, if there is to be state intervention in this fundamentally private process, then a responsibility arises to provide the treated infant with a living environment that is able to accommodate his or her special needs. This is what the state would expect of parents who decided to consent to aggressive treatment. Therefore, a similar infant whose guardianship is taken over by the state for reasons of natural parent "neglect" should be afforded no less an opportunity to thrive and develop.

It is precisely the intrusive nature of the state's *"en parens patriae"* proceedings in the severely impaired critically ill infant decision-making dilemma that gives rise to a responsibility to provide adequately for such an infant's special needs, lest the state become a "neglectful guardian" *after* aggressive medical intervention has saved the life of the infant. To provide simply custodial care thereafter to such an infant whose continued existence is the direct result of state authority neither serves the best interests of the child, nor fulfills the guardianship responsibility that was supplanted from the natural parents. Government exercise of such intrusive power must not be irresponsible or purposeless.

The standard proposed is not unreasonably difficult, nor is it arbitrarily drawn for severely impaired newborns. I contend that it is the same standard that exists for critically ill *un*impaired newborns. As discussed earlier, there is clear precedent for state intrusion into medical decision making when parents would deny medical treatment to a critically ill nonimpaired infant/child. I submit that government authority to intervene in this context is predicated on the knowledge that such nonimpaired children will be readily adoptable to loving and caring parents who will provide the proper adoptive home environment to satisfy the best interests of the child. It is this outcome-oriented principle of responsible post-illness guardianship that is the essential aspect of my concerns. If finding willing prospective parents who are capable of providing a proper environment for special needs children is difficult, then the fact of such difficulty ought not to lessen the responsibility of state authorities. To accept a standard that provides less care/responsibility for special needs children would be discriminatory against such infants and neglectful of their best interests.

References

Berseth, C. L. (1987) Ethical dilemmas in the neonatal intensive care unit. *Mayo Clin. Proc.* **62,** 67.

Bowen v. American Hospital Association, et al. (1986) US Supreme Court. 106, S. Ct. 2101, 2113, 2123.

Caulfield, P. K. (1986) The Child Abuse Amendments of 1984: Inadequate procedural due process safeguards. *Valparaiso University Law Rev.* **2,** 103,124.

House Select Committee on Children, Youth and Families (1987) Abused Children in America: Victims of Official Neglect, H.R. Doc. No. 260, 100th Cong., 1st Sess., p. 43.

Huefner, D. S. (1986) Severely handicapped infants with life-threatening conditions: Federal intrusions into the decision not to treat. *Am. J. Law Med.* **12,** 188.

I.A.C.M. § 19-371: Health Care Facilities—Duties of Hospitals (effective 7/1/87). Maryland is the only state to have legislated this requirement.

Kopelman, L. M., Irons, T. G., and Kopelman, A. E. (1988) Neonatologists judge the "Baby Doe" regulations. *N. Engl. J. Med.* **318,** 677,680,683.

Lantos, J. (1987) Sounding board—Baby Doe five years later. *N. Engl. J. Med.* **317,** 444.

Murray, T. H. (1985) The final anti-climactic rule on Baby Doe. *Hastings Cent. Rep.* **5,** 15.

National Right to Life News (1985) **7,** 1.

Newman, S. A. (1989) Baby Doe, Congress and the states: Challenging the federal treatment standard for impaired infants, *Am. J. Law Med.* **15,** 1,4.

Office of Inspector General, US Department of Health and Human Surveys (1987) *Survey of State Baby Doe Programs* 11. *Supra*, p. 33.

President's Commission for the Study of Ethical Problems in Medicine and Biomedical and Behavioral Research (1983) *Deciding to Forego Life-Sustaining Treatment: Ethical, Medical and Legal Issues in Treatment Decisions* (US Government Printing Office, Washington, DC).

Rhoden, N. (1985) Treatment dilemmas for imperiled newborns: Why quality of life counts, 58 S. *California Law Rev.* **58,** 1283, 1292.

Time Nobody's children. (Oct. 9, 1989) p. 91.

42 USC § 5102 (1984).

Appendix

*Chronology of Events Related to Passage
of the 1984 Child Abuse Amendments*

April 9, 1982. Baby Doe was born in Bloomington, Indiana, suffering from Down syndrome, an esophageal atresia, and a tracheoesophageal fistula.

May 18, 1982. DHHS published Notice to Health Care Providers: "Discriminating Against Handicapped by Withholding Treatment or Nourishment" (Federal Register Vol. 47, No. 116). The notice, based on Section 504 of 1973 Rehabilitation Act, advised hospitals receiving federal funds that it was unlawful to withhold nutritional, medical, or surgical treatment from an infant with a disability if the withholding was based on the fact that the infant was handicapped, and the handicap did not render the treatment or nutritional sustenance medically contraindicated.

May 26, 1982. Representative John Erlenborn introduced legislation in Congress to prevent denial of treatment to children with disabilities. This legislation was later modified and eventually became the Child Abuse Amendments of 1984.

March 7, 1983. DHHS published its Interim Final Rule: "Nondiscrimination on the Basis of Handicap" (Federal Register, Vol. 48, No. 45). It was based on Section 504 of the 1973 Rehabilitation Act and provided that hospitals receiving federal aid would be required to post notices describing the protections of federal law against discrimination toward the handicapped and listing a toll free federal "hotline" telephone number so that violations could be reported. It also provided for onsite hospital investigations by DHHS personnel with 24-hour access to patient medical records.

March 1983. President's Commission for the Study of Ethical Problems in Medicine and Biomedical and Behavioral Research issued its report, entitled *Deciding to Forego Life-Sustaining Treatment.*

April 14, 1983. American Academy of Pediatrics v. Heckler 561 F. Supp. 395 (1983). Federal District Court for the District of Columbia invalidated the Interim Final Rule on the grounds that DHSS had not followed appropriate procedural requirements because the rule was not published for public comment.

October 11, 1983. Baby Jane Doe was born in Port Jefferson, New York, suffering from spina bifida, hydrocephalus, microcephaly, and related complications.

January 12, 1984. DHHS published its Final Rule: "Nondiscrimination on the Basis of Handicap: Procedures and Guidelines Relating to Health Care for Handicapped Infants" (Federal Register, Vol. 49, No. 8). This rule was also based on Section 504 of the 1973 Rehabilitation Act and required the posting of notices stating that failure to feed and care for handicapped infants was a violation of federal law and listing DHHS and state child protective services telephone numbers. It allowed for onsite visits with 24-hour access to medical records, and required state CPS agencies receiving federal child abuse monies to establish and maintain procedures to prevent instances of medical neglect. It encouraged, but did not mandate, the creation of Infant Care Review Committees and allowed health care workers to use reasonable medical judgment in selecting alternative courses of treatment. Futile treatment or treatment that would merely temporarily prolong the act of dying were not considered medically beneficial.

October 9, 1984. President Reagan signed Public Law 98-457, the Child Abuse Amendments of 1984. The statute required that states establish programs for the reporting of medical neglect (including instances of withholding of medically indicated treatment from disabled infants with life-threatening conditions) as a condition of receiving federal funds for their child protective services systems.

April 15, 1985. DHHS issued its Final Rule to implement the Child Abuse Amendments of 1984: "Child Abuse and Neglect Prevention and Treatment Program" (Federal Register, Vol. 50, No. 72). The Final Rule established procedures and guidelines for implementing the Child Abuse Amendments of 1984. It defined terms used in the statute, explained the role of state child protective services agencies, outlined the procedures for notification in cases of medical neglect, and provided that state child protective services agencies should pursue legal proceedings as necessary. The Final Rule included an appendix containing interpretative guidelines that discussed in detail the meaning of the language used in the statute. Model guidelines for establishing Infant Care Review Committees were published simultaneously. Establishment of such committees was encouraged, but not mandated.

June 9, 1986. Bowen v. American Hospital Association, 106 S. Ct. 2101 (1986). The US Supreme Court upheld a lower federal court decision invalidating DHHS's January 12, 1984 Final Rule entitled "Nondiscrimination on the Basis of Handicap: Procedures and Guidelines Relating to Health Care for Handicapped Infants" (based on Section 504 of the 1973 Rehabilitation Act) on the grounds that: 1. The hospital's withholding of treatment when no parental consent had been given did not violate the 1973 Rehabilitation Act; 2. The regulations, which required posting, reporting, and access to records, were not founded on evidence of discrimination and were totally foreign to the authority conferred on the Secretary of DHSS; 3. The Secretary could not dispense with the Act's focus on discrimination, and instead, employ federal resources to save the lives of handicapped newborns without regard to whether or not they were victims of discrimination by hospitals receiving federal funds.

    October 25, 1991. President Bush signed Public Law 101-126, reauthorizing the Child Abuse Amendments of 1984.

Biographies

Patricia Barber, Ph.D., is an Assistant Research Scientist at the University of Kansas, a Research Associate with the Beach Center on Families and Disability, and Courtesy Professor in Special Education. Her professional interests include family policy, disability policy, and research and theory related to children with disabilities or chronic illnesses and their families. She is coauthor of *A Community Approach to an Integrated Service System for Children with Special Needs* (Paul H. Brookes, 1988).

Robert H. Blank, Ph.D., is a Political Scientist with affiliations at the University of Canterbury in Christchurch, New Zealand, and Northern Illinois University. He has published 15 books and numerous articles and book chapters, primarily in the area of biomedical policy. Among his most recent books are *Rationing Medicine* (Columbia, 1988), *Life, Death, and Public Policy* (Northern Illinois, 1988), and *Regulating Reproduction* (Columbia, 1990). He is a member of the US Congress Office of the Technology Assessment Advisory Panel on Neuroscience Research and has lectured widely on policy issues in medicine. Current projects include books on reproductive rights, fertility control, and workplace hazards.

James Bopp, Jr., J.D., is partner in the law firm of Brames, McCormick, Bopp and Abel, in Terre Haute, Indiana. He is President of the National Legal Center for the Medically Dependent and Disabled, Inc., and General Counsel for the National Right to Life Committee, Inc. He has served on a number of government committees and panels, including the President's Committee on Mental Retardation, the NIH Human Fetal Tissue Transplantation Research Panel, and the Congressional Biomedical Ethics

Advisory Committee. He has written many articles and book chapters, edited two books, entitled *Human Life and Health Care Ethics* (University Publications of America, 1985) and *Restoring the Right to Life: The Human Life Amendments* (Brigham Young, 1984), and is currently an editor for *Issues in Law & Medicine*.

A. G. M. Campbell, M.B., F.R.C.P., received his early training in medicine and pediatrics in Scotland and London, England. In 1962–1963, he was an Assistant Chief Resident at the Children's Hospital of Philadelphia and in 1963–1964 was a Fellow in Pediatric Cardiology at the Hospital for Sick Children in Toronto. After two years doing fetal and neonatal research at the Nuffield Institute for Medical Research in Oxford, he returned to the United States in 1967 as an Assistant Professor of Pediatrics at the Yale University School of Medicine in New Haven, Connecticut. In 1968 he became Associate Professor and Director of the Newborn Services at the Yale-New Haven Hospital. In 1973 he returned to Scotland as Professor of Child Health and Pediatrics at the University of Aberdeen and Consultant Pediatrician at the Royal Aberdeen Children's Hospital.

Arthur L. Caplan, Ph.D., is Director of the Center for Biomedical Ethics, Professor of Surgery, and Professor of Philosophy at the University of Minnesota. He is the former Associate Director of the Hastings Center, and has taught at Columbia University's College of Physicians and Surgeons and at the University of Pittsburgh. He is the author or editor of 12 books, including *Scientific Controversies* (Cambridge, 1987), *Which Babies Shall Live?* (Humana, 1985), *The Sociobiology Debate* (Harper and Row, 1978), *Concepts of Health and Disease* (Addison-Wesley, 1981), and *In Search of Equity* (Plenum, 1983). He has also published more than 200 articles. He has written for many newspapers, including the *Washington Post, The New York Times,* and the *Los Angeles Times,* and served as a consultant to many organizations, including the New York Academy of Sciences, the New Jersey Department of Health, the Minnesota Department of Health, the Office of Technology Assessment of the US Congress, the National Institutes of Health, the National Endowment for the Humanities, and the Institute of Medicine of the National Academy of Sciences.

Norman Fost, M.D., M.P.H., is Professor and Vice Chairman of Pediatrics and Director of the Program in Medical Ethics at the University of Wisconsin School of Medicine. At Wisconsin, he is also Director of the Residency Training Program, Coordinator of the Child Protection Team, Chair of the Hospital Ethics Committee, and Chair of the Institutional Review Board. He is past Chair of the American Academy of Pediatrics Committee on Bioethics and the author of the A.A.P.'s Guidelines for Infant Bioethics Committees. He is also a Fellow of the Hastings Center and the author of numerous publications on ethical and legal issues in health care, particularly involving children.

Thomas G. Irons, M.D., is Professor of Pediatrics and Associate Dean at East Carolina University School of Medicine. He has published articles in the *American Journal of Medical Genetics,* the *New England Journal of Medicine,* the *Archives of Internal Medicine,* the *Journal of General Internal Medicine,* and *Contemporary Pediatrics,* and has lectured widely, usually on issues related to children.

Arthur E. Kopelman, M.D., is Professor of Pediatrics and Section Head of Neonatology at East Carolina University School of Medicine. His research interests relate to the prevention of neurologic injury in premature infants. He has published articles in the *Journal of Pediatrics,* the *New England Journal of Medicine,* and the *American Journal of Diseases of Children.*

Loretta M. Kopelman, Ph.D., is Professor and Chair of the Department of Medical Humanities at East Carolina University School of Medicine. She has coedited three anthologies published by Kluwer: *Ethics and Mental Retardation* (1984), *Ethics and Critical Care Medicine* (1985), and *Children and Health Care* (1989). She has published in the *New England Journal of Medicine,* the *Journal of the American Medical Association,* the *Journal of Philosophy and Medicine,* and other journals. She serves on the editorial board of *The Encyclopedia of Bioethics* (2nd edition) and *The Journal of Medicine and Philosophy.*

Janet Marquis, Ph.D., is a Research Statistician with the Schiefelbusch Institute of Life Span Studies at the University of Kansas. She holds degrees in mathematics, education, computer science, and educational research. Her professional interests include research design, multivariate data analysis, and information management. She is coeditor of *Cognitive Coping Research and Developmental Disabilities,* which will be published by Paul H. Brookes.

Janna C. Merrick, Ph.D., is Professor and Chair of the Department of Political Science at St. Cloud State University. She is a former Visiting Scholar at the Hastings Center and a former Visiting Scholar at the Center for Biomedical Ethics, University of Minnesota School of Medicine. Her research area is health care policy with a focus on issues relating to pregnancy and infants. She has published in *Politics and the Life Sciences,* the *Policy Studies Review,* and in several anthologies. She serves on the editorial board of *Politics and the Life Sciences* and is coediting special symposia for the *Journal of Legal Medicine* and for *Women and Politics.*

Mary Nimz, J.D., is Chief Staff Counsel for the National Legal Center for the Medically Dependent & Disabled, Inc. She has represented amicus clients in cases involving withdrawal of nutrition and hydration, and she recently represented several infants born with spina bifida, along with their parents, alleging discriminatory denial of treatment against a hospital. She writes in the area of disability rights policy and has recently published in *Clearinghouse Review* and *Issues in Law and Medicine*. She also serves as a contributing editor to *Issues in Law and Medicine*.

Joy Penticuff, Ph.D., is Associate Professor at the University of Texas at Austin School of Nursing. She teaches health policy, health care ethics, and clinical courses in high-risk perinatal nursing. She is a Fellow of the American Academy of Nursing, a former Visiting Scholar at the Hastings Center, and a member of its Research Group on Ethics and the Care of the Newborn. Her most recent research is in perinatal ethics, conducted through an NIH National Research Service Award Senior Fellowship at the Center for Ethics, Baylor College of Medicine, Houston, Texas. Dr. Penticuff's clinical specialization is neonatal intensive care nursing.

Anthony Shaw, M.D., was educated at Harvard College and New York University School of Medicine with training in general and pediatric surgery at Columbia-Presbyterian Medical Center in New York City. He is Professor of Surgery at UCLA and Chief of Pediatric Surgery at Los Angeles County-Olive View Medical Center in Sylmar, California. He has practiced and taught pediatric surgery for over thirty years. Dr. Shaw has been chair of the Ethics Committee of UCLA Medical Center and the Ethics Committee of the American Pediatric Surgical Association. For the past eight years, he has served as the liaison member of the American Academy of Pediatrics Committee on Bioethics from the surgical section of the A.A.P. His writings in the early 1970s were among the first to bring "Baby Doe" issues to public attention.

Rud Turnbull, LL.B., LL.M., is Professor of Special Education, Courtesy Professor of Law, and Codirector of the Beach Center on Families and Disability at the University of Kansas. He specializes in disability law, special education law, public policy analysis, and applied ethics in the field of disability. He has held national elected offices in the American Association on Mental Retardation, the Association for Retarded Citizens of the United States, and the Association for Persons with Severe Disabilities. He has been a Public Policy Fellow of the Jos. P. Kennedy, Jr. Foundation, and in that role Special Counsel to the United States Senate Subcommittee on Disability Policy.

Terry Walman, M.D., J.D., is both a physician and an attorney. He has practiced medicine for 14 years, and he is board-certified in Anesthesiology and in Critical Care Medicine. Dr. Walman is also admitted to the Bar of Maryland and is a Fellow of the American College of Legal Medicine. He is a member of the faculty of the Johns Hopkins University School of Medicine.

Robert F. Weir, Ph.D., is currently the Director of the Program in Biomedical Ethics at the University of Iowa College of Medicine in Iowa City. He is also a Professor in the Department of Pediatrics and in the School of Religion. He is the author of *Selective Nontreatment of Handicapped Newborns* (Oxford, 1984) as well as several articles on ethical issues in neonatalogy. In addition, he has edited *Ethical Issues in Death and Dying* (Columbia, 1977, 1986) and has written *Abating Treatment with Critically Ill Patients* (Oxford, 1988).

Index

Index

Actual person, 10–12
Administrative Proc-
    edures Act (APA),
    75
Allen, William B., 35, 214
American Academy
    of Pediatrics (AAP),
    42, 46–50, 59, 66,
    69, 77, 79, 80, 89,
    164, 190, 191, 197,
    238, 239, 288, 299
*American Academy
    of Pediatrics v.
    Heckler*, 47, 53, 74,
    76
American Association of
    Medical Colleges, 79
American Association
    of Mental Defi-
    ciency, 69, 89
American Civil Liberties
    Union, 44, 109
American College of Obste-
    tricians and Gyne-
    cologists, 48, 55,
    79, 89
American College
    of Physicians, 79, 89
American Hospital Assoc-
    iation, 48, 55, 79
American Institutional
    Review Boards, 227
*American Journal of Law
    and Medicine*, 201

American Life Lobby,
    52, 53, 56, 59
American Medical Associa-
    tion (AMA), 55, 56,
    61, 86, 89, 110, 252
American Nurses Associa-
    tion, 24, 69, 89
Americans United for Life,
    59
Anderson, James E., 37
Arthur, Leonard, 108
Association for Persons
    with Severe Handi-
    caps, 69, 79
Association for Retarded
    Citizens, 52, 56, 69,
    79, 89, 197

Baby boy G, 6, 7
Baby Doc controversy,
    105–122, 226, 255,
    286
Baby Doe debate, 35–72
"Baby Doe" regulations,
    3, 14, 46, 66, 67,
    105, 106, 112–114,
    121, 164, 176, 192,
    194, 195, 197, 202,
    237–266, 287, 289,
    292, 294, 305
"Baby Doe Special
    Assignment," 77
Baby girl A, 3, 4, 13
Baby girl L, 5, 6

Baby girl T, 7
Baby Howle, 192
Baby Jane Doe, 43, 52, 53,
    64, 75, 76, 107, 111,
    115, 120, 123, 143,
    195, 262, 309
Baby Lance, 264
Barber v. People, 82
Bartlett v. Glendale
    Adventists Med.
    Center, 193
BDR-I, 238, 239, 253, 256,
    257, 260, 261
BDR-II, 238, 239, 242–253,
    255–265
Beneficence, 210
Benjamin, Martin, 18
Best-interests position
    (standard), 17–22,
    25, 30
Bleich, David J., 189
Bloomington Hospital, 36
Bowen v. Am. Hosp. Ass'n.,
    76, 105, 124, 158,
    195, 237, 238,
    253–258, 261–263,
    301–304
Boyle, Joseph, 56
British Hospital Ethics
    Advisory Commit-
    tees, 227
British National Health
    Service, 226
Broody, Howard, 18
Bush administration, 26

Campbell, A. G. M., 40, 63,
    116, 119, 186, 187,
    207–236
Cardiopulmonary resusci-
    tation (CPR), 171,
    172
Caufield, Patricia, 300
Chicago Tribune, 163
1984 Child Abuse Amend-
    ments, 26, 57–62,
    64–67, 73–103, 105,
    106, 112, 124, 130,
    148, 151, 164, 265,
    267–284, 299–316
Child Abuse and Neglect
    Prevention and
    Treatment Act, 198,
    199, 238
Child Abuse Prevention
    and Treatment Act,
    54, 57
Child Abuse Prevention
    and Treatment Act
    Program (CAPTA),
    309–311
Child abuse regulations,
    3, 4, 12, 14, 23, 26
Child Protective Service
    System, 198
Child Protective Services
    (CPS), 46, 51, 53,
    54, 58, 59, 62, 63,
    94, 96, 100, 262
Child Protective Services
    Agencies, 227

Children's Hospital
National Medical
Center, 47, 48
Christian Action Council,
48, 49
Christian Scientists, 202
Cobb, Roger W., 38
Columbia University's
Medical Center, 119
Commission Report, 94–96
Committee for Child
Protection, 196
Congress, 36–40, 45, 48,
53, 54, 56, 67, 76,
81, 83, 111, 124,
212, 300, 301
Cranston, Alan, 54, 69,
78, 80
Cruzan v. Director, 152

"Death in the Nursery,"
43, 46
Department of Education,
50
Department of Health
and Human
Services (DHS,
DHHS), 4, 23, 40,
43, 44, 46–51, 53,
54, 57, 59, 61, 62,
66, 74, 75, 77, 81,
83, 84, 86–88, 90,
93, 94, 96–99, 105,
106, 111, 116, 124,
194, 212, 237, 238,
240, 260, 262, 287,

289–291, 302, 304
Department of Justice, 44,
46, 53
Destro, Robert, 214
Diagnosis-related group
(DRG), 27
Doctor–patient relation-
ship, 287
Doe v. Bloomington Hosp.,
1983
Do-not-resuscitate status,
7
Down syndrome, 36, 41–43,
49, 63, 64, 73, 107,
108, 111–114, 116–
120, 186, 188, 193,
200, 218–222, 290,
292, 306, 310
Duff, Raymond S., 40, 43,
63, 116, 119, 186,
187, 211, 277

Edelman, Murray, 37, 45,
48, 49, 52, 56, 68
Elder, Charles D., 38
Erlenborn, John, 76

Federal Child Abuse Act,
287
Feinberg, Joel
"Common sense
personhood," 9, 10
fetal interests, 19
future interests, 19
Final Rule, 75, 82, 300
Fletcher, Anne, 24

Fletcher, Joseph, 187
Fost, Norman, 42, 63,
        285–297
Frader, Joel, 24

Gesell, Gerhard, 194, 261

Haggerty, Robert J., 164
Hastings Center,
        Kennedy Institute,
        187, 192
Hastings Center Report,
        58, 188, 197
Hatch, Orrin, 54, 69, 82
Hospital–patient
        relationships, 47
House, the, 57, 69
HR, 1904, 57

Illinois Department
        of Children and
        Family Services, 42
Indiana Supreme Court,
        36, 74
Infant Care Review
        Committee (ICRC),
        4, 13, 49–51, 59, 66,
        95–98, 100, 198, 199,
        227, 268, 272, 273,
        280, 285–297, 305
In re Baby Girl Muller,
        91, 92
In re Conroy, 82
In re Dinnerstein, 193
In re Quinlan, 192, 286
In re Spring, 193

In re Steinhaus, 90
Institute of Medicine's
        Committee to Study
        Outreach for Prena-
        tal Care, 175
Institute of Medicine's
        Committee to Study
        the Prevention of
        Low Birth-Weight,
        166
Institutional Review
        Boards (IRBs), 290,
        291, 295
Interim Final Rule, 61, 74
Intermittent Positive
        Pressure Ventila-
        tion (IPPV), 222,
        224, 225
Interpretative Guidelines,
        82–84, 86, 88

Johns Hopkins University,
        42, 186
Joint Explanatory
        Statement, 82, 88
Joseph P. Kennedy
        Foundation, 42
*Journal of Pediatric
        Surgery,* 191

Kennedy, Edward, 57, 77,
        78, 80, 187
Koop, C. Everett, 14, 48,
        49, 51, 53, 56, 211
Kopelman Study,
        65, 67, 305

Lesch-Nyhan's Syndrome,
117
Life and death decisions,
1–33
Life-sustaining interventions, 5
Life-sustaining technologies, 23
Life sustaining treatment,
12–18, 21–23, 25–30
forgoing, 207–236
Likert scales, 128
Limitation of Treatment
Bill, 219, 232
Linares, Sammy, 4, 5, 14

MacNeal Hospital, 4
Medicaid, 26, 48
Medicare, 48
Moskop, John, 58
Muller, Rebecca Jean,
91, 92
Murray, Thomas, 10
Myelomeningocele, 125

National Association
of Children's Hospitals and Related
Institutions, 47, 55,
69
National Center for the
Medically Dependent and Disabled, 44
National Commission
to Prevent Infant
Mortality, 175
National Down Syndrome
Congress, 56, 69, 79,
89, 197
National Health Service,
229, 230
National Institute
for Disability and
Rehabilitation
Research, 123
National Right to Life
Committee, 52, 56,
69, 79, 89, 299
National intensive care,
208–210
Neonatal Intensive Care
Unit (NICU), 1–3,
6–8, 12, 18, 21, 24–
30, 43, 44, 66, 68,
147, 155–183, 199,
203, 248, 249, 251,
255, 258, 259, 263,
265, 268–273, 275–
281, 299–316
*New England Journal
of Medicine*, 40, 62,
116, 186, 190, 265
Newmann, Steven, 311
New York Court
of Appeals, 52
New York Times Magazine,
185
Nonmaleficence, 210
Nutrition and hydration,

22–25, 36, 81, 82, 84, 85

Office for Special Education and Rehabilitative Services, 50
Office of Civil Rights, 61
Office of Health Economics, 159
Office of Inspector General (OIG), 254
Office of the Inspector General of the Department of Health and Human Services, 113
Office of Technology Assessment, 27
Oklahoma Children's Memorial Hospital, 63

Parham v. J. R., 151, 152
Paris, John, 24
Patient's best interests (*see also* best-interest position [standard]), 5, 6, 18, 21, 22, 25, 29
Pauly, Mark, 229
Pediatric Intensive Care Unit (PICU), 4, 5, 8, 26
Pediatrics, 191
Physician–patient

relationship, 308
Pomerance, Jeffrey J., 66
Post-Baby Doe period, 40
Potential person, 10, 11
President's Commission on the Ethical and Legal Problems in Medicine and Biomedical and Behavioral Research (President's Commission), 47, 48, 50, 55, 61, 164, 188, 201, 287, 304, 308
Principles of Treatment of Disabled Infants, 124, 151

Quality of life, advocates, 17
assessment, 80, 195
concerns, 166
considerations, 52
criteria, 12
decision making, 77
distinctions, 176
judgments, 95
position, 17
potential, 165
sanctity-of-life debate, 165
Quinlan, Karen, 286

Ramsey, Paul, 187
Reagan administration,

14, 26, 36, 39, 45, 48, 52, 60, 195, 301, 302
"re B minor," 108, 115
1973 Rehabilitation Act, 26, 44, 46, 53, 54, 61, 74, 82, 83, 88, 95, 99, 109, 112, 124, 194, 195, 212, 238, 302
Rex Trailer Co. v. United States, 99
Rhoden, Nancy, 311
Roe v. Wade, 9
Rush-Presbyterian-St. Luke's Medical Center, 4

Saldanha, Rita, 58
Senate, the, 57, 69, 82
Senate Committee on Labor and Human Resources, 77, 78
Separation-of-powers principle, 37
Shaw, Anthony, 41, 42, 185–205
Singer, Peter, 15
Spina bifida, 43, 44, 111, 113, 114, 116–120, 123–153, 187, 292
Spina Bifida Association of America, 69, 79, 89, 197
Stanford Law Review, 193

Steinhaus, Lance, 89–91, 264
Stevens, John Paul, 43
Suffolk County Supreme Court, 52
Superior Court of Monroe County, 36
1987 Survey of State Baby Doe Programs, 62

Tay-Sachs, 117, 311
Tooley, Michael, 15
Trisomy 13, 218, 220, 241, 242, 310
Trisomy 18, 218, 220, 310
Trisomy 21, 306

United States Court of Appeals, 75
United States v. Marion County School District, 99
United States v. University Hosp., 53, 75, 76
University of Kansas, 123
University of Oklahoma Health Sciences Center, 42
US Civil Rights Commission, 35, 63–65, 94, 95, 97, 195, 196, 199, 201, 203, 212–215, 290
US Constitution, 9, 37, 67
US Department of Health, Education, and

Welfare (HEW),
286, 290
US Supreme Court, 36, 53,
54, 112, 124, 151,
152, 158, 195, 238,
253, 254, 257, 261–
263, 301, 303, 304

VATER syndrome, 7
Very low-birthweight
(VLBW) babies,
167–172, 175,
177–179, 208, 209,
223, 225
Village Voice, 196

Ward, Justice, 220
Warren, Mary Anne, 15
Washington Post, 54
Wexler, Leonard D., 261
"Who Lives, Who Dies?," 44
Wills, George, 43, 56
Women's and Infants
Hospital of Rhode
Island, 169

Yale–New Haven Hospital,
40, 116
Yale New Haven Special
Care Nursery, 187